CURTISS ARMY H... in action

by Larry Davis

Color by Don Greer & Tom Tullis
Illustrated by Joe Sewell

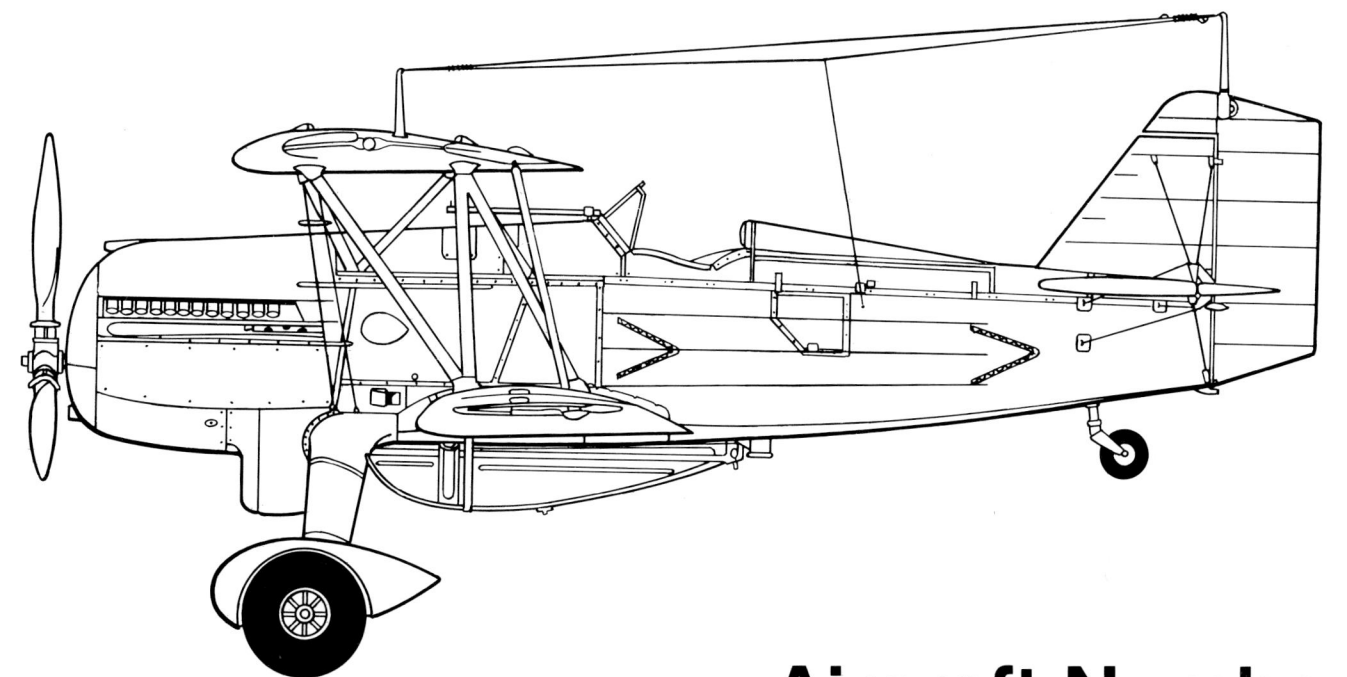

Aircraft Number 128
squadron/signal publications

A 17th Pursuit Squadron P-6E Hawk and an A-3B Attack Falcon of the 90th Attack Squadron form up during joint Army maneuvers held in 1933.

COPYRIGHT © 1992 SQUADRON/SIGNAL PUBLICATIONS, INC.
1115 CROWLEY DRIVE   CARROLLTON, TEXAS 75011-5010
All rights reserved. No part of this publication may be reproduced, stored in a retrieval system or transmitted in any form by any means electrical, mechanical or otherwise, without written permission of the publisher.

ISBN 0-89747-286-1

If you have any photographs of the aircraft, armor, soldiers or ships of any nation, particularly wartime snapshots, why not share them with us and help make Squadron/Signal's books all the more interesting and complete in the future. Any photograph sent to us will be copied and the original returned. The donor will be fully credited for any photos used. Please send them to:

Squadron/Signal Publications, Inc.
1115 Crowley Drive.
Carrollton, TX 75011-5010.

Acknowledgements

Air Force Museum
Jack Binder
Peter Bowers
Jeff Ethell
Steven Hudek

Vincent Berinati
Warren Bodie
Robert Esposito
Don Garrett, Jr.
Nicholas J. Waters III

Dedication

To the memory of one of the true aviation pioneers - Glenn Hammond Curtiss.

Six P-6Es of the 17th Pursuit Squadron fly formation above a solid undercast. The command stripes painted on the upper wing indicated the pilot's position within the group — three stripes for the group commander, two for the squadron commander and one stripe indicated a flight or element leader. (AFM)

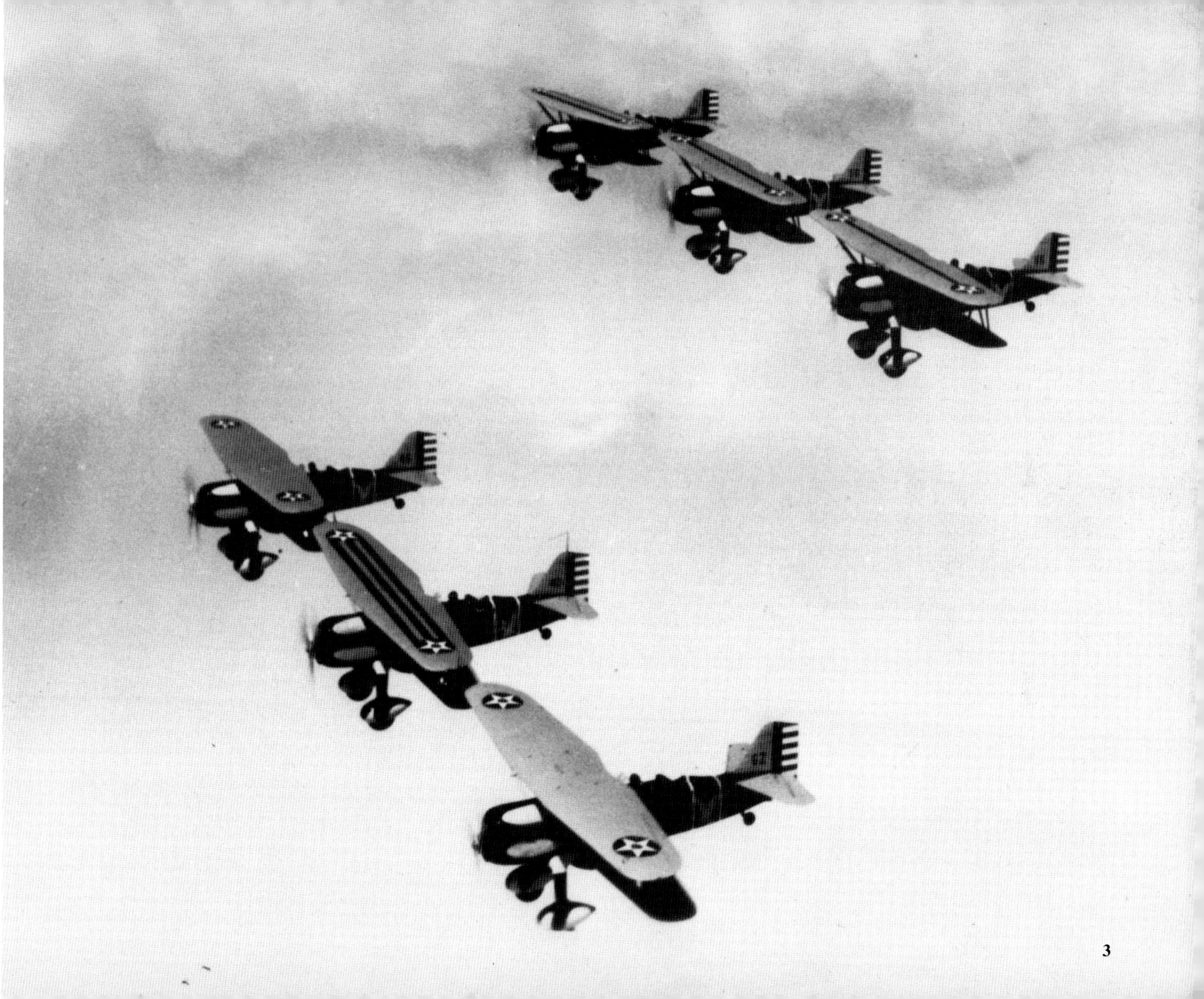

Introduction

With aviation enthusiasts, some things never change. Mention jet fighters and (depending on the time period) enthusiasts will automatically think of — Eagles, Phantoms or Sabres. Talk about propeller driven fighters and you'll hear about Mustangs, Spitfires, and Messerschmitts. But bring up the subject of U.S. Army Air Corps biplane "pursuits" and the conversation will usually turn to Curtiss Hawks. For it was the Curtiss Hawk, more than any other Army aircraft, that captured the imagination of the public during the 1920s and 1930s.

The Hawk wasn't the fastest or most maneuverable fighter of the period, but it did have something no other pursuit ship of the time had — personality! Plain and simple, it "looked" right. It had the sleek, aggressive look that the public felt a pursuit ship should have. The Hawks, and their Falcon derivatives, kept the Curtiss Aeroplane and Motor Company solvent through the Depression and lean budgets of the 1930s. This, in turn, enabled Curtiss to build the more potent Hawks that fought on all fronts in the Second World War.

Few people realize the impact that Glenn Hammond Curtiss had on aviation history. Although the Wright Brothers are credited with the invention of the airplane, Glenn Curtiss is credited with several of the significant developments that led to the modern military aircraft. As a member of Alexander Graham Bell's Aerial Experiment Association, Curtiss developed engines to be used for various flying machines. His own aircraft, named *June Bug*, first flew on 21 June 1908. Later that same year Curtiss modified the aircraft with a set of pontoons, making it the first attempt at a seaplane. In August of 1909 the Curtiss *Golden Flyer* beat the Wright Brothers entry in the Gorden Bennett trophy race. In June of 1910 Curtiss conducted the first successful aerial attack against a military warship; although, the significance of this event (and Billy Mitchell's later demonstrations) would be ignored until 7 December 1941.

On 14 November 1910, Glenn Curtiss made an even more significant demonstration when the *Golden Flyer* made the first successful launch from a U.S. warship. The first

Army LT Russell Maughan won the 1922 Pulitzer Trophy Race flying the #2 Curtiss R-6 racing/pursuit aircraft at a speed of 248.5 mph. The aircraft was powered by a 460 hp Curtiss D12 liquid cooled engine. (AFM)

The PW-8 was the first production pursuit aircraft built by Curtiss. PW-8 stood for Pursuit, Water-cooled design #8. The XPW-8 prototype rolled out in January 1923 and was fitted with flush-mounted radiators built into the upper wing surface. (AFM)

deck landing came a year later when the aircraft landed on the aft deck of USS PENNSYLVANIA. This landing also marked the first use of arresting gear on a naval vessel. The *Golden Flyer*, equipped with pontoons, became the first successful seaplane when it broke water on San Diego Bay on 26 January 1911. This same aircraft (fitted with a wheel/float combination) became the first amphibious aircraft.

The first U.S. Navy aircraft was the Curtiss A.1. Other Curtiss "firsts" include: the first passenger seaplane takeoff and landing, first aviator license, first enclosed hull on a seaplane, first use of a Vee-hull on a seaplane and the first company to mass-produce an aircraft (five flying boats for the Navy and three more for the Army).

The First World War saw the Curtiss Aeroplane and Motor Company move to the forefront of aircraft production, not only in the United States but the world! Curtiss built flying boats for both the U.S. and Royal Navy, and license-built several foreign designs for use by the U.S. military. By 1917, orders for Curtiss aircraft outgrew the company's production facilities and other manufacturers like Boeing and Lockheed license-built several Curtiss designs to meet the demand.

It was the Curtiss Model JN, the immortal Jenny, that helped move Curtiss to the forefront of aircraft production. The Curtiss JN-4 Jenny served as a trainer aircraft for the U.S., British and Canadian Air Services, with more than 10,000 being built. After the end of the "war to end all wars," Jenny's went on to be stars of the barnstorming era. It was the Jenny that brought flying to every little farm town across the U.S.

During the early 1920s Curtiss turned toward the construction of high speed pursuit and racing aircraft. In March of 1922 the Navy commissioned the USS LANGLEY as its first aircraft carrier. In May of that same year, Curtiss unveiled the first fighter aircraft specifically designed to operate from an aircraft carrier — the TS-1.

In the early 1920s, Curtiss aircraft dominated the air racing scene. Curtiss combined several design characteristics to produce a pair of fighter/racer aircraft designs: the CF-1 and CF-2, later designated CR-1 and CR-2. These used wooden fuselages developed along the lines of Curtiss' successful seaplane designs. The sleek, highly polished racers, combined with a new Curtiss D-12 power plant, easily won the 1921 Pulitzer Trophy Race for the Navy at an average speed of 176.7 mph.

By 1922, a keen rivalry had developed between the Army and Navy racing teams. Curtiss modified the 1921 Pulitzer Trophy winning aircraft with a new engine as the Navy's entry. The Army, however, wanted a new aircraft. To meet the Army requirement, Curtiss created the R-6. Although similar to the Navy CRs, the R-6 was much smaller and over 600 pounds lighter. The end of the 1922 Pulitzer Trophy Race saw the two Army R-6s in 1st and 2nd place, while the Navy CRs placed 3rd and 4th (out of a sixteen ship field).

Ski-equipped PW-8s of the 1st Pursuit Group prepare for a flight from Selfridge Field, Michigan, to Oscoda, Michigan during 1925. The bulges in the upper wing are water tanks for the flush-mounted radiators. The wing radiators were a constant source of trouble. (AFM)

Adding insult to injury, Army LT Russell Maughan set a new Absolute World Speed Record in an R-6 at 248.5 mph. Curtiss racers dominated the air racing scene and it wasn't until 1926 that a serious challenger (from Europe) emerged. That race was the last to see a pure racing design entered under the Curtiss name.

During 1923, Curtiss began work on the aircraft that led to the Curtiss Hawk family. The new pursuit ship was a direct development of the Curtiss R2C/R3C racing aircraft program. Powered by a 440 hp Curtiss D-12 engine, the XPW-8 (Experimental Pursuit Water-cooled design number 8) first flew during January of 1923. It was quite similar in appearance to the racers, but the XPW-8 was much larger and the wing was in a standard position above the upper fuselage. The radiators were radical in design in that they were flush-mounted on the upper wing surface directly in front of the pilot to reduce drag. Radiator and welding technology being what it was in the early 1920s, these flush-mounted radiators were a constant source of problems. They leaked often, dousing the pilot with boiling water.

Being impressed with the performance of the Curtiss racers, and recognizing the close tie between the new pursuit ship and the racers, the Army ordered twenty-five aircraft under the designation PW-8 in September of 1923. LT Russell Maughan used one of the first production PW-8s to set a new trans-continental speed record during June of 1924.

A new company was fast emerging in the military aircraft field — Boeing. In April of 1923 Boeing unveiled the XPW-9. Powered by the same Curtiss D-12 engine as the PW-8, the Boeing pursuit was much lighter, had the wing leading and trailing edges tapered and had the radiator mounted vertically under the nose in a tunnel. A flyoff between the XPW-9 and a production PW-8 during 1924 resulted in a clear win for the Boeing aircraft. It was more maneuverable and had a greater rate of climb, mostly due to its lighter weight. As a result, Boeing was given a contract to build thirty PW-9s.

Curtiss immediately adopted the tunnel radiator in its follow-on aircraft, the XPW-8A.

A PW-8 of the 27th Pursuit Squadron at the 1924 Dayton Air Race. At this time the aircraft of the Army Air Service were painted overall Olive Drab with vertical Red, White and Blue rudder stripes. (AFM)

The production PW-8A was powered by the same Curtiss D-12 rated at 440 hp. The top speed was 178 mph at sea level, with a rate of climb of 1,830 ft/min and a service ceiling of 22,000 feet. Armament was the same as the PW-8: a pair of .30 caliber machine guns mounted above the engine in the cowling. The weight of the PW-8A was 2,830 pounds, some 300 pounds lighter than the PW-8 with its flush-mounted radiators.

To improve the aircraft's maneuverability, Curtiss realized that the answer was to use the Boeing-developed tapered wing. Using photographs of the Boeing wing as a model, Curtiss engineers came up with a tapered wing of their own. By December of 1924, Curtiss had modified the XPW-8A with a short-span, tapered wing. The results of flight tests with the XPW-8B were very positive and the Army ordered the new tapered-wing design into production. One other change would take place prior to production. During 1926, the U.S. Army Air Service became the U.S. Army Air Corps and all aircraft designations were changed. The first aircraft ordered into production for the new Army Air Corps was the production variant of the XPW-8B, now designated the Curtiss P-1, nicknamed the Hawk.

LT Russell Maughan set a new trans-continental speed record of 21 hours 58 minutes on 23 June 1924 in this PW-8. It marked the first time that the continental United States had been crossed in less than a day. (AFM)

Development

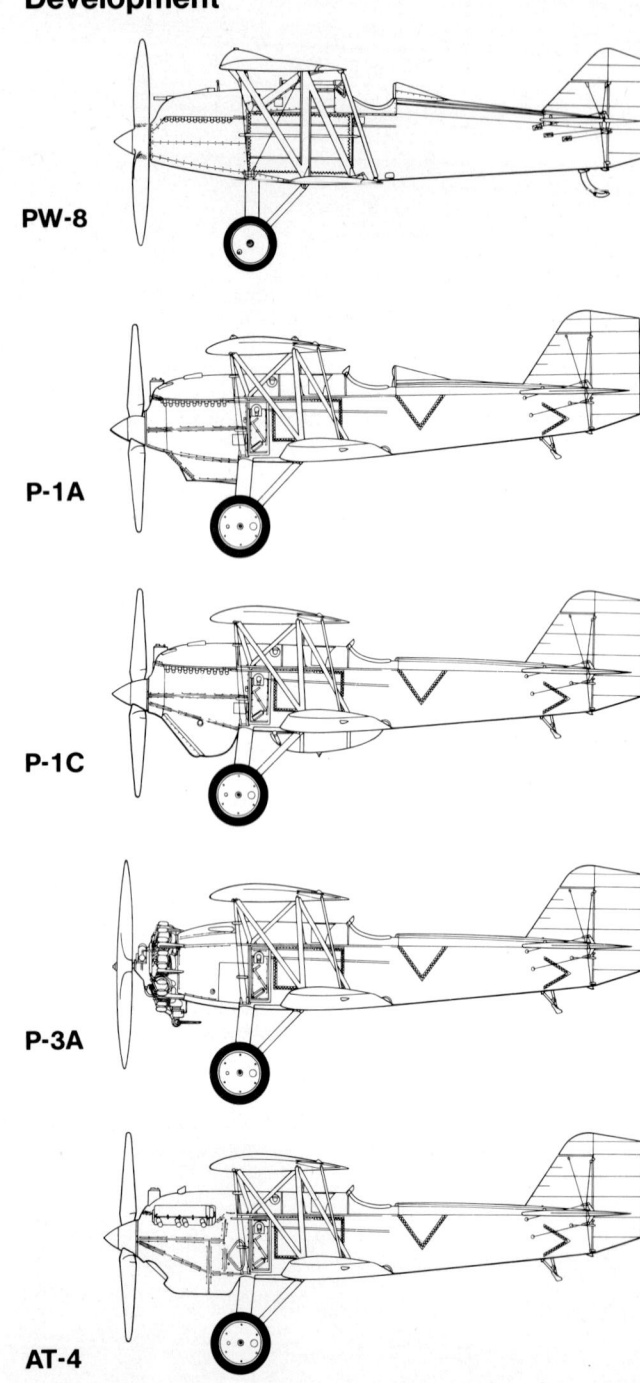

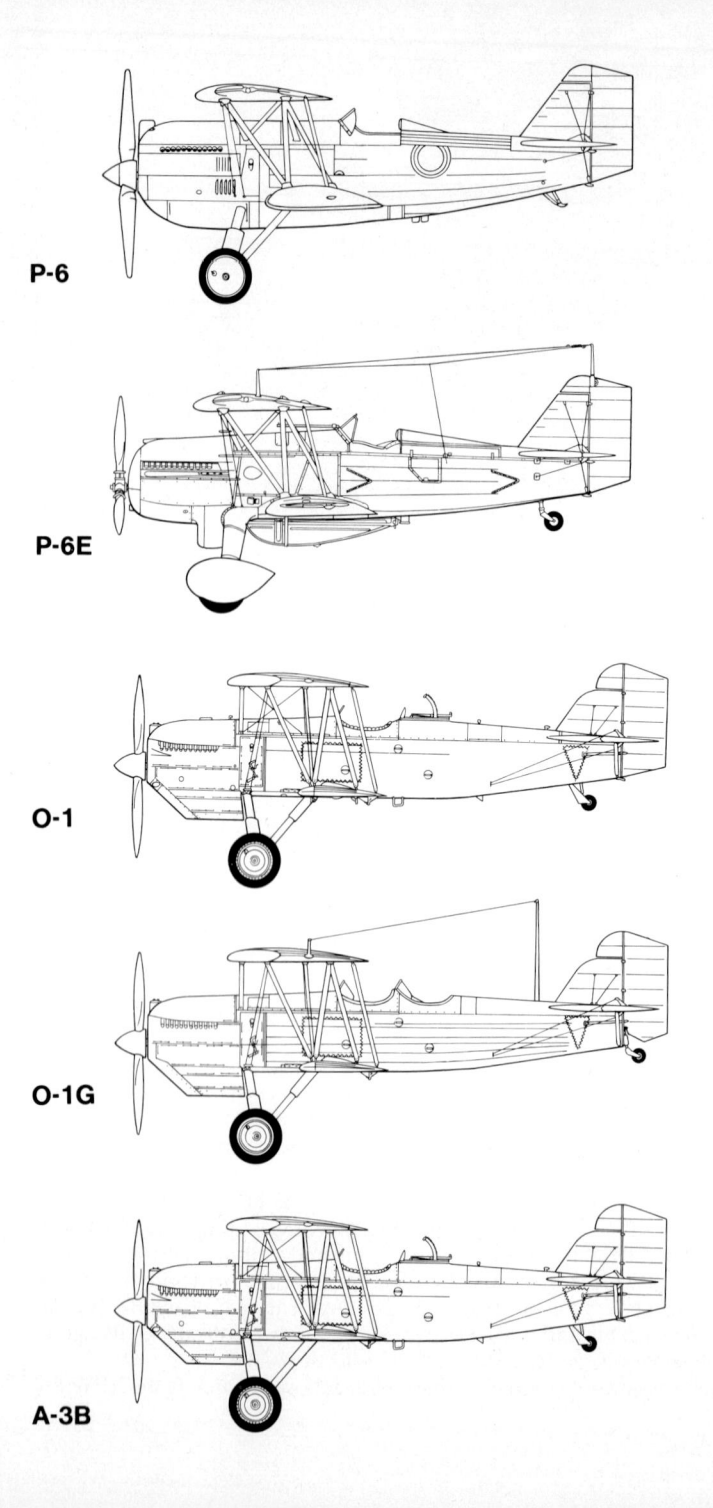

PW-8

P-1A

P-1C

P-3A

AT-4

P-6

P-6E

O-1

O-1G

A-3B

P-1 Hawk

The Army issued Curtiss a contract in 1926 to build fifteen production variants of the XPW-8B under the designation P-1. The aircraft was given the trade name "Hawk" by Curtiss and this soon caught on, becoming the popular name for the aircraft as well. When the Army Air Service became the Army Air Corps in 1926, the PW designation was dropped with all fighter-type aircraft becoming simply P (for Pursuit). The only difference between the prototype XPW-8B and the P-1 was in the rudder balance area. The P-1 was powered by an 1,150 cubic inch, 435 hp, Curtiss D-12 in-line, liquid cooled engine (Army designation V-1150-1) giving the aircraft a top speed of 163 mph at sea level, a rate of climb of 1,800 ft/min and a service ceiling of 22,000 feet.

The armament consisted of a pair of .30 caliber machine guns mounted in the upper nose and firing through the propeller arc. The aircraft could also be configured with either a fifty gallon jettisonable under fuselage fuel tank or a 200 pound bomb load. With a full load of fuel and ammunition, the aircraft had a gross weight of 2,846 pounds.

Although the Army ordered fifteen P-1 Hawks, only ten were delivered as P-1s (serials 25-410 through 25-419), the last five aircraft on the contract being completed as P-2 Hawks. Navy interest in the Army Hawk resulted in a contract for nine aircraft under the designation F6C-1, which were literally copies of the Army P-1 Hawk. The first production P-1 Hawk was delivered to the Air Service test center at McCook Field in July of 1925 and the first operational aircraft went to the 1st Pursuit Group at Selfridge Field (near Detroit), Michigan in October of 1925.

P-1A

The P-1A differed from the P-1 in having the internal fuselage structure lengthened three inches, resulting in an overall increase in the aircraft's length of one inch. Additionally, the fuel system was improved, as was the bomb release mechanism. The Army ordered twenty-five of these improved Hawks on 9 September 1925. Curtiss delivered the first production P-1A to the 1st Pursuit Group at Selfridge in April of 1926. Only twenty-three P-1As were delivered from the original contract, since two aircraft were pulled off the Curtiss assembly line and finished as the AT-4 Advanced Trainer prototype and P-3 prototype.

P-1B

The P-1B incorporated a new engine, the improved Curtiss V-1150-3 in-line engine, which was a more reliable power plant but offered no increase in available horsepower. The P-1B also had the radiator and radiator housing redesigned, larger wheels and improved engine/flight control systems. These changes increased the empty weight by almost 100 pounds. With no increase in available engine power, the top speed of the P-1B fell to 157 mph at sea level. More importantly, the rate of climb dropped from 1,810 ft/min to 1,540 ft/min. The Army ordered twenty-five P-1Bs in August of 1926 with the first production aircraft being delivered in November of 1926 to the Air Corps Detachment at Bolling Field, Washington, D.C.

A Boeing PW-9C from the Bolling Field Air Detachment, Washington, D.C. The Boeing PW-9 introduced both the under fuselage, tunnel-mounted radiator and the tapered wing into service. Both these features were adapted by Curtiss for the Hawk series. (AFM)

Fuselage Development

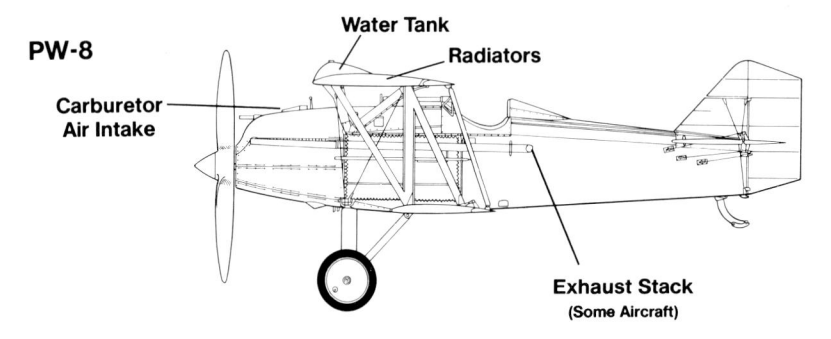

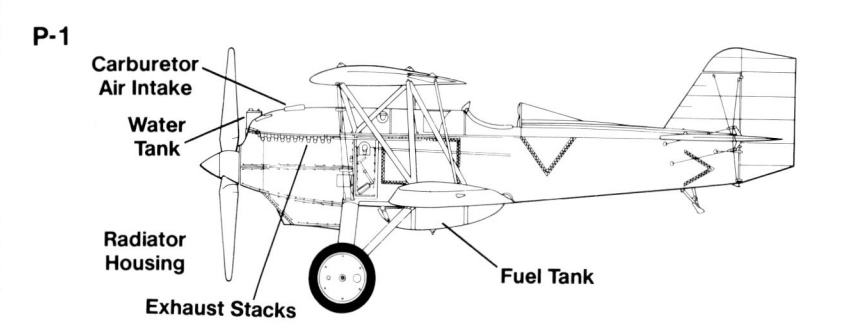

PW-8

P-1

P-1C

The P-1C used the V-1150-5 engine, which made improvements in engine reliability, with no increase in horse power. The P-1C had a narrower nose section, tapered radiator, larger and heavier wheels and tires and mechanical brakes. Empty weight increased another 50 pounds and with this increase the P-1C's top speed decreased to 154 mph, while the rate of climb fell to 1,460 ft/min. The Army ordered thirty-three P-1Cs in October of 1928 with the first aircraft being delivered to the 1st Pursuit Group at Selfridge in January of 1929.

AT-4/P-1D

The Army decided that a first line pursuit aircraft fitted with a lower horsepower engine would be ideal to fill the advanced trainer role. As a result of this decision, a P-1A airframe was pulled off the Curtiss assembly line and re-engined with a 180 hp Hispano "E" engine, license-built by Wright under the designation V-720. The aircraft was a standard P-1A airframe, without weapons, and the reduced power resulted in a top speed of 132 mph.

Even with this poor performance, the Army ordered forty of these advanced trainer aircraft, under the designation AT-4. With the exception of the engine and the lack of armament, the AT-4 was identical to the P-1A.

The Army soon realized that the AT-4 was totally unsuitable for the trainer role. Production aircraft began to reach the school units at Kelly Field in May of 1927 and it was quickly obvious that the performance of the AT-4 was miserably lacking in all areas. In fact, the aircraft was quite dangerous in the hands of a student pilot. In 1928, the Army ordered thirty-five out of the forty AT-4s re-engined with the 435 hp Curtiss V-1150-3 engine (the same engine used in the P-1B). This brought performance up to a level equal to that of the other Hawk pursuits then in service. The re-engined AT-4s, armed with a single .30 caliber machine gun, were re-designated as P-1Ds.

AT-5/P-1E

The last five AT-4 aircraft were refitted with 220 hp Wright R-790-1 radial engines and re-designated as AT-5s. Although the available power was increased by forty horse power over the Wright 'E' engine, the performance of the AT-5 actually fell off even more, with the top speed falling to 125 mph. The Army had Curtiss re-engine four of the AT-5s with 435 hp V-1150-3 engines, re-designating the aircraft as P-1Es.

AT-5A/P-1F

The P-1F was a similar conversion of some thirty-five production AT-5A advanced trainers. The AT-5A was powered by a 220 hp R-790-1 radial engine and its airframe had been brought up to P-1B standards with larger wheels, brakes and improved engine controls. With this added weight, the performance of the AT-5A suffered, with the top speed dropping to 122 mph. Deliveries of the AT-5A to Kelly Field school squadrons began in June of 1928. During 1929, however, the Army ordered the thirty-one AT-5As that had been delivered re-engined with the 435 hp V-1150-3 engine and armed with one .30 cal. gun. Once the conversion was complete, the aircraft were re-designated as P-1Fs.

P-2

The last five aircraft on the original Army P-1 contract were built with highly modified engines under the designation P-2 Hawk. The V-1150-1 engines were bored out to a displacement of 1,400 cubic inches and fitted with superchargers giving the engine a rating of 505 hp. With this increase in power, the P-2 had a top speed of 172 mph and a rate of climb of 2,170 ft/min. The supercharger gave the P-2 a service ceiling of 24,000 feet; however, problems with the new supercharger cancelled out the advantages of the increase in performance. As a result, the Army had three of the P-2s re-engined with standard V-1150-1 engines and redesignated them as P-1As. One of the remaining P-2s later became the prototype for the P-6 series.

P-3

The last P-1A from the initial production run was modified to accept a new 410 hp

The Curtiss 435 hp D-12 (V-1150-1) water cooled, inline engine, powered both the P-1 Hawk series and the later O-1 Falcon series of aircraft. The oval object in the front of the engine was the water tank for the radiator and the oval intake in the middle was the carburetor air intake. (AFM)

An overall Olive Drab P-1A Hawk of the 17th Pursuit Squadron based at Selfridge Field, Michigan during 1924. The P-1 Hawk featured a tapered wing and tunnel radiator. It was the first Curtiss fighter to carry the trade name Hawk, which later became accepted as the popular name for the aircraft as well. (AFM)

Pratt & Whitney R-1340 'Wasp' air-cooled radial engine. Designated the XP-3A, the radial-engine Hawk was used as an engine test bed throughout its service life. The combination of the big air-cooled radial engine and Hawk airframe would later prove to be a very good one — for the Navy! The Army was committed to the use of the liquid-cooled, heavier V-12 engine in its Hawks. The simpler and easier to maintain Pratt & Whitney Wasp radial was perfect for the Navy and they procured a large number of radial engined Curtiss Hawks for the fleet.

The Army did purchase five P-3A Hawks during 1928. The basic airframe was the same as the P-1B/AT-5A with larger wheels, brakes and improved flight controls. Powered by the 410 hp Pratt & Whitney R-1340-3 Wasp, the P-3A Hawk had a top speed of 154 mph and a rate of climb of 1,742 ft/min, neither of which was better than the D-12 powered Hawk then in service. The service ceiling did improve by almost 2,000 feet to 23,000 feet. But Army decided against use of the radial engine for its Hawks. The P-3As did see operational service with all five being delivered to the 1st Pursuit Group at Selfridge Field in October of 1928.

P-5

There were five P-5 Hawks built by Curtiss during 1928. These aircraft were basically P-1Bs powered by a 435 hp turbo-supercharged Curtiss D-12F engine (V-1150-3). The addition of the turbo-supercharger brought the gross weight of the P-5 up to 3,349 pounds, some 500 pounds more than a P-1 Hawk. Additionally, the turbo-supercharger added nothing but drag at lower altitudes. The sea level top speed of the P-5 was 142 mph, but the top speed at altitude rose significantly to a rather phenomenal 173 mph at 25,000 feet. The rate of climb was poor, being only 1,150 ft/min, although the service ceiling greatly improved. The P-5 had a ceiling of 31,000 feet, more than 10,000 feet above the P-1A. Clearly the turbo-supercharger offered significant improvements in performance at altitude, although its numerous technical problems killed the P-5 and the program was halted after the delivery of the five production aircraft in 1928.

Air Corps Cadet Wiseley crashed this P-1A of the 43rd School Squadron at Kelly Field, Texas during 1928. The aircraft was painted to simulate a First World War German Fokker fighter for the movie 'Wings.' (AFM)

This 95th PS P-1 Hawk has the wings and tail surfaces painted in a high visibility Yellow #4 (FS 13432) (commonly called Chrome Yellow) scheme, which was standardized by Army in 1927. At this same time, the rudder stripes were changed from vertical to horizontal. (AFM)

Nose Development

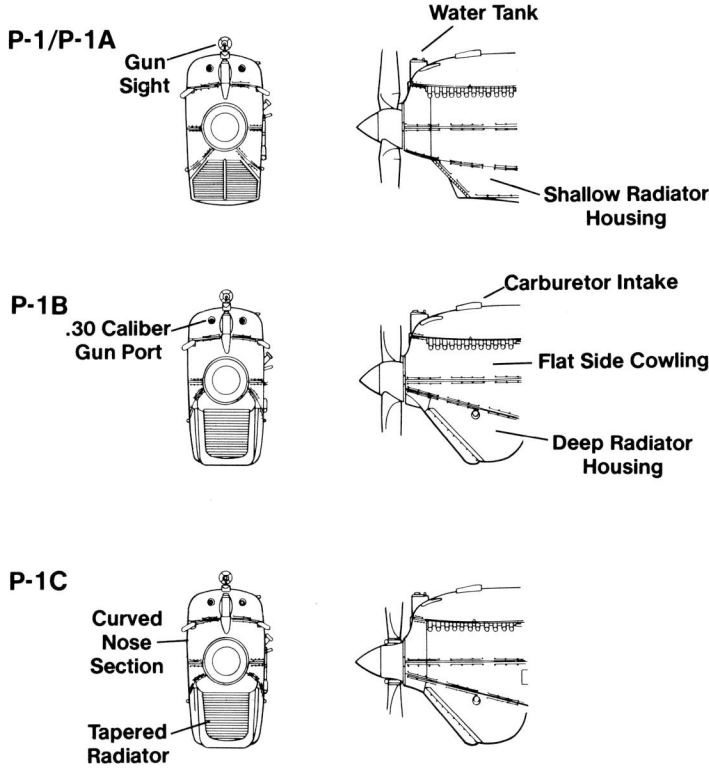

P-1/P-1A — Gun Sight, Water Tank, Shallow Radiator Housing

P-1B — .30 Caliber Gun Port, Carburetor Intake, Flat Side Cowling, Deep Radiator Housing

P-1C — Curved Nose Section, Tapered Radiator

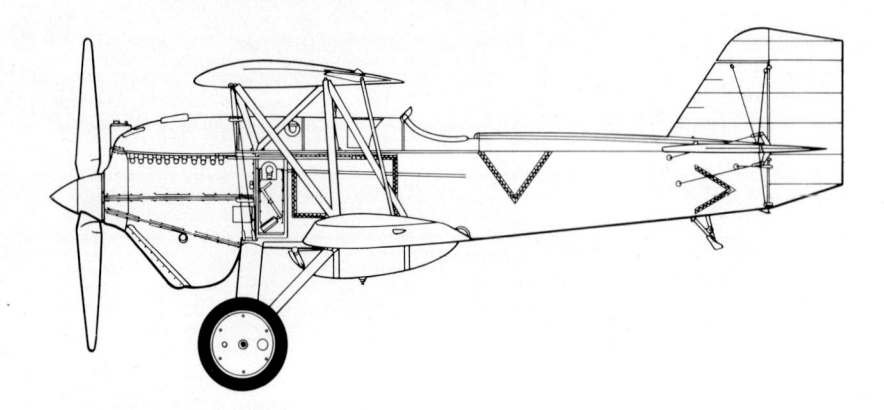

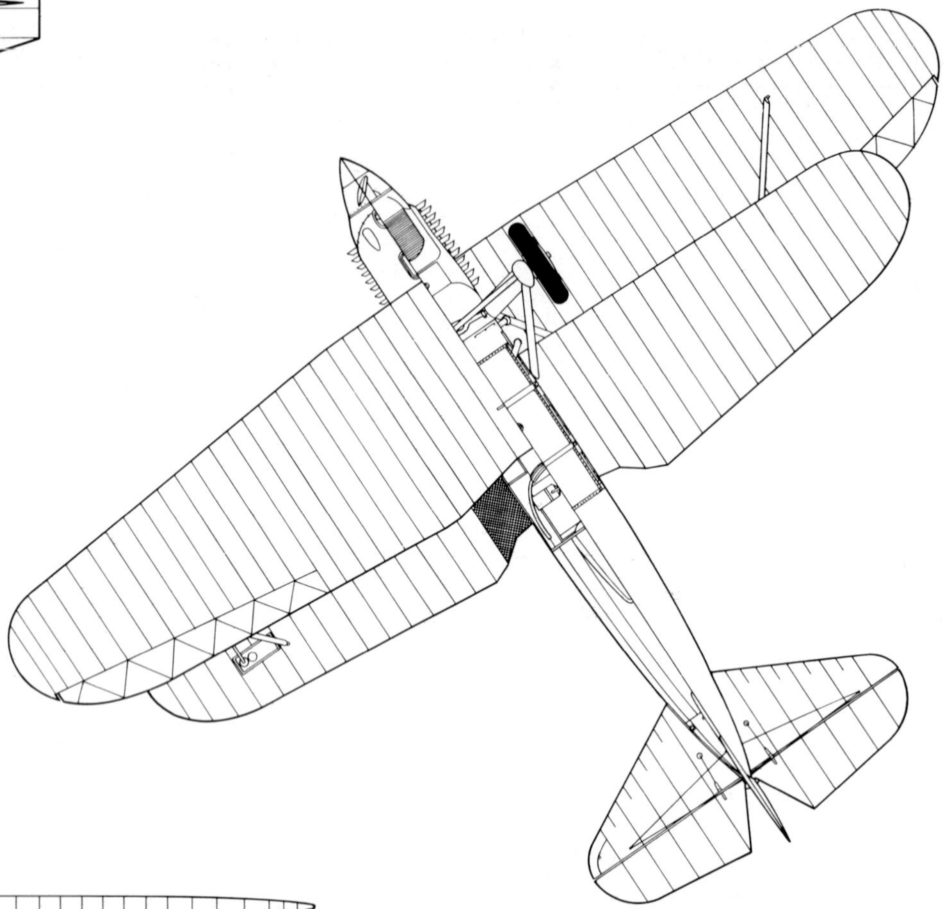

Specifications

Curtiss P-1C Hawk

Wingspan	31 feet 6 inches
Length	23 feet
Height	8 feet 9 inches
Empty Weight	2,195 pounds
Maximum Weight	2,973 pounds
Powerplants	One 435 hp Curtiss D-12C liquid cooled engine.
Armament	Two .30 caliber machine guns

Performance
  Maximum Speed ....154.4 mph
  Service ceiling ....20,800 feet
  Range ....300 miles
Crew ....One

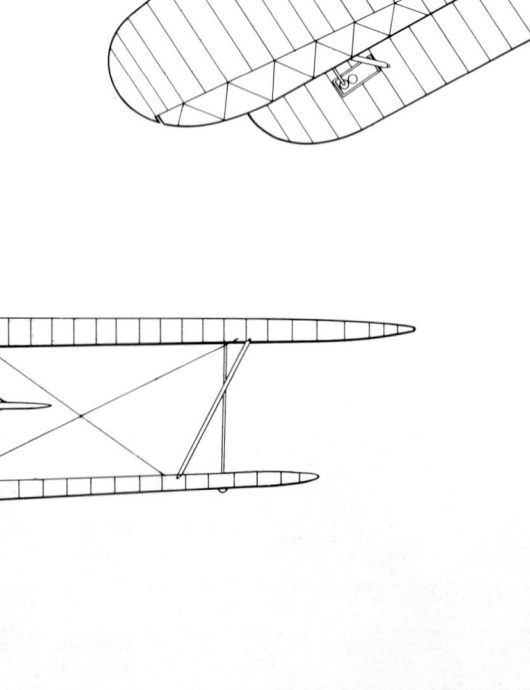

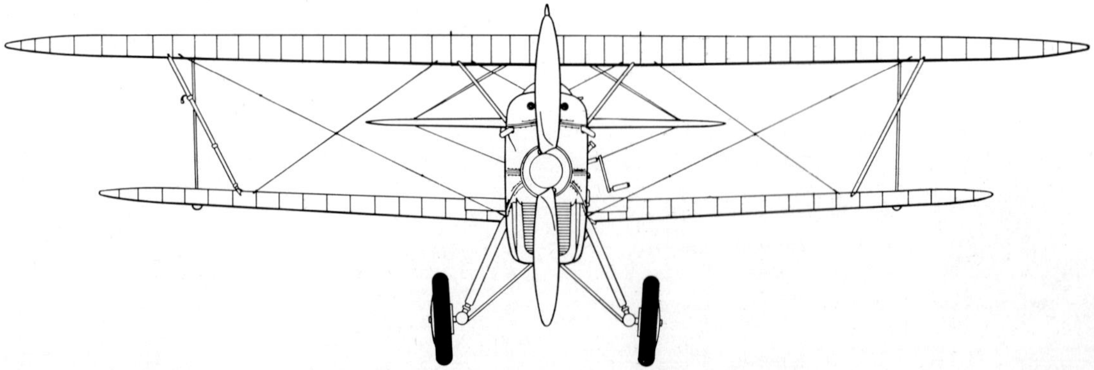

10

A P-1A Hawk of the 43rd School Squadron at Kelly Field, San Antonio, Texas. The V-shaped panel under the cockpit was a laced up inspection panel in the fabric covered fuselage. (Warren Bodie)

A ski-equipped P-1B Hawk of the 27th Pursuit Squadron on the Selfridge ramp during the Winter of 1927. The P-1B had a much deeper radiator housing than the P-1A although the radiator itself was narrower. The nose was Red with a thin White outline. (AFM)

An Army Air Corps pilot runs up the engine of the first P-1B (27-63) on the snow-covered ramp at Bolling Field, Washington. The insignia is that of the Air Corps Detachment at Bolling Field, which was later adopted by the 14th Bomb Squadron. (Warren Bodie)

11

This P-1B Hawk carries a mixture of standardized paint schemes. It had an overall Olive Drab fuselage and wings with "U.S. Army" in White under the wing, but also had the later style horizontal rudder stripes. The aircraft carries a fifty-six gallon fuel tank under the fuselage. (AFM)

A P-1B of the 1st Pursuit Group at Selfridge Field, Michigan during 1928. The P-1B had an improved V-1150-3 engine and larger wheels than the P-1A. The aircraft was Olive Drab with Yellow wings/tail and had the 1st PG insignia painted on the fuselage under the cockpit. (AFM)

A pair of P-1C Hawks of 43rd School Squadron on a gravel airstrip near Kelly Field, Texas during 1928. The P-1C had larger wheels and mechanical brakes. Both of these aircraft have wire mesh mud and stone guards over the main wheels. (AFM)

The P-1C was powered by a 435 hp V-1150-5 engine, which had improved reliability over the earlier engines, but no increase in rated horsepower. The radiator was tapered with a more rounded lower opening. (AFM)

MAJ Clarence Tinker flew this P-1D while assigned to the 43rd School Squadron at Kelly Field. P-1Ds were re-engined AT-4 advanced trainers. The aircraft were powered by a 435 hp V-1150-3 engine in place of the 180 hp Hispano "E" used on the AT-4. The armament consisted of a single .30 caliber machine gun in the cowl. (AFM)

A ski equipped P-1C of the 1st Pursuit Group parked in the snow and ice near the Selfridge ramp during the Winter of 1929. The skis were mounted directly to the landing gear wheel spindles and could be easily replaced by standard wheel landing gear at any time. (AFM)

The AT-4 was a standard P-1A airframe powered by a 180 hp Wright "E" (Hispano) engine. The AT-4 was so underpowered that it was considered very dangerous to fly and most were later re-engined with Curtiss D-12 engines and re-designated as P-1Ds. (AFM)

A pair of AT-4s from the Air Corps Technical School at Kelly Field. The Army ordered forty AT-4s, completing thirty-five of them before issuing orders that they would all be re-engined with D-12 engines as P-1Ds. (Ed Huebner)

The AT-5 series were powered by the 220 hp Wright J-5 Whirlwind air cooled radial engine. Still underpowered, the AT-5 and AT-5A were both re-engined with Curtiss D-12 engines and redesignated as P-1Es. A total of thirty-one AT-5As were built. (AFM)

Nose Development

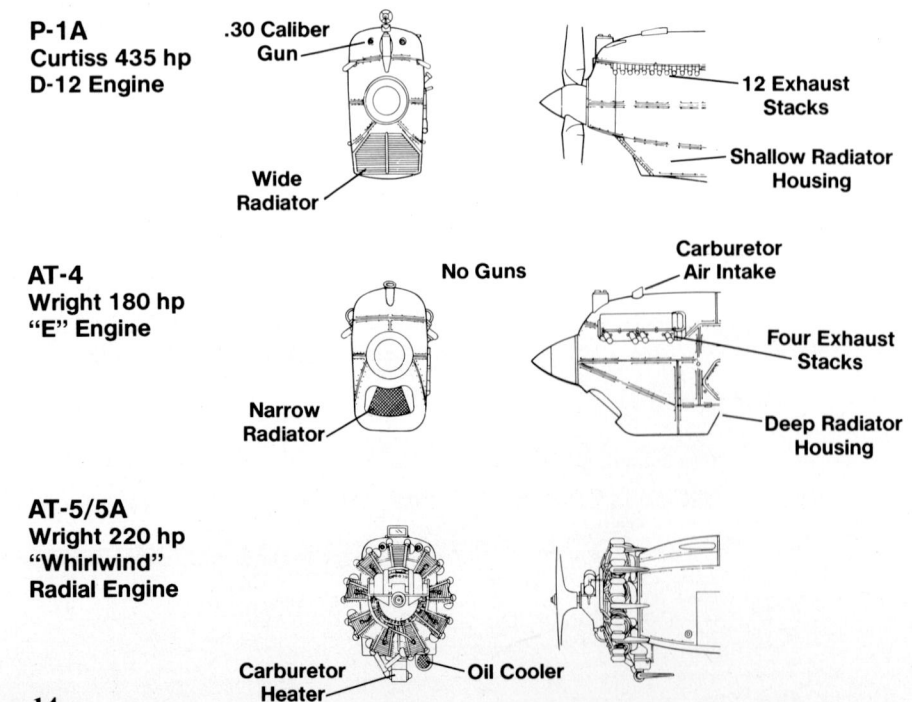

P-1A
Curtiss 435 hp
D-12 Engine
- .30 Caliber Gun
- Wide Radiator
- 12 Exhaust Stacks
- Shallow Radiator Housing

AT-4
Wright 180 hp
"E" Engine
- No Guns
- Narrow Radiator
- Carburetor Air Intake
- Four Exhaust Stacks
- Deep Radiator Housing

AT-5/5A
Wright 220 hp
"Whirlwind"
Radial Engine
- Carburetor Heater
- Oil Cooler

The Curtiss P-3 Hawk is generally considered the "father" of the Navy and export Hawk series. The P-3A was a P-1A airframe powered by a 410 hp Pratt & Whitney R-1340-1 Wasp air cooled radial engine. The Army bought five P-3s but the Navy ordered thirty-one aircraft under the designation F6C-4. (AFM)

XP-6

The prototype XP-6 Hawk started life as the fourth production P-2 airframe. The aircraft was returned to the Curtiss Garden City, New Jersey plant and re-engined with the new 600 hp liquid cooled in-line Curtiss V-1570 "Conqueror" engine to allow the aircraft to compete in the 1927 National Air Races. The contract for the XP-6 prototype was let in April of 1927, although work actually started some weeks earlier.

When the aircraft emerged. it was similar in appearance to all the previous Hawk variants; in fact, very little had been changed to the rear of the fuselage engine firewall. The changes were all centered around the engine and radiator installation. The XP-6 was entered in the 1927 National Air Race, placing second to what amounted to a purpose built racing aircraft. The XP-6, flown by LT AJ Lyon, averaged 189 mph while the purpose built racing aircraft that beat the XP-6 averaged 201 mph. That aircraft was the Curtiss XP-6A.

After the race, the XP-6 was returned to Wright Field and assigned the test number XP-494. Re-engined with a 525 hp V-1570-1 engine and fitted with military equipment, the aircraft was evaluated as a new pursuit aircraft. The XP-6 had a top speed of 176 mph, a rate of climb of 2,277 ft/min and a service ceiling of 23,110 feet.

The winner of the race, the XP-6A, was built using several different Curtiss aircraft as a basis. The fuselage was that of a P-1A Hawk which was re-engined with one of the new 600 hp Conqueror engines, boosted to some 730 hp for racing. The engine was cowled in a PW-8 engine cowling and the tunnel radiator and its housing was discarded in favor of the untapered wings and flush-mounted radiators of the XPW-8A. The result was an aircraft that was aerodynamically smoother than the other Hawks. Powered by the higher horsepower Conqueror engine, the XP-6A, flown by LT Earl C. Batten easily ran away from its 1927 competition becoming the first U.S. Army Air Corps aircraft to break the 200 mph barrier (average speed 201 mph).

P-6

Satisfied with the results of the 1927 National Air Race and the showing made by a virtually combat ready aircraft, the XP-6, the Army ordered the P-6 into production on 3 October 1928. Powered by the 600 hp V-1570-17 engine, the P-6 looked like a P-1A with a deeper, more rounded nose. This change in profile was because the Conqueror was a larger engine and required a larger radiator for engine cooling. The Army tied future development of the P-6 with the development of a new cooling liquid called Prestone coolant. Prestone (ethylene glycol) was a better cooling agent than water because it had both a lower freezing and higher boiling point than water. Prestone development, however, was lagging and the Army decided to order the P-6 into production using a standard water radiator, which resulted in the deep rounded forward fuselage contours.

The P-6 also differed from the earlier P-1 series in the landing gear. The P-6 used a strengthened landing gear with an oleo-pneumatic shock absorber mounted in the forward landing gear strut. This landing gear had been tested on the last two production P-1Cs and was selected as standard equipment for the P-6 series.

Performance of the P-6 closely matched that of the XP-6 and the Army ordered a total of eighteen aircraft. The first production aircraft were delivered to the Wright Field Test Center at Dayton, Ohio, in October of 1929. Operations with the 1st Pursuit Group at Selfridge Field began a month later, in November of 1929.

The Army halted P-6 production after ten aircraft, ordering that the remaining eight

The XP-1B was an engine test bed to conduct flight tests of new power plants. It was fitted with a small, experimental radiator which was filled with a new coolant called Ethylene Glycol, commonly called Prestone. Prestone radiator development ultimately led to the P-6E Hawk. (AFM)

The first P-6 Hawk on the grass at Selfridge Field, Michigan during 1929, shortly after it was delivered to the 27th Pursuit Squadron. The aircraft was powered by a 600 hp Curtiss V-1570 engine. (AFM)

aircraft on the production order be completed with Prestone coolant radiators in place of the standard water radiator. Being a better coolant than water, less Prestone was required for engine cooling and as a result, a smaller radiator was needed on a Prestone-equipped aircraft. The Prestone-cooled aircraft retained the deep radiator housing of the P-6, even though they were equipped with a much smaller radiator. The last eight production aircraft with the Prestone radiators were designated as P-6As and delivered to the 1st Pursuit Group.

The XP-6B was a one-off aircraft built specifically for long range record flights. The XP-6B was a P-1C airframe (29-529) re-engined with a V-1570-1 engine. The radiator housing resembled the housing on the P-6 but was much rounder in appearance. The biggest change was in the internal fuel capacity which was increased from 50 gallons to 250 gallons. CAPT Ross Hoyt flew the XP-6B named *Hoyt Special* from Mitchell Field near New York City to Nome, Alaska in thirty-four hours and twenty minutes (air time) on 18/19 July 1929. The XP-6B crashed during the return flight and was trucked back to Wright Field. After being repaired, the aircraft was retained at Wright Field for test work, being retired in August of 1931. No production aircraft carried the designations P-6B or P-6C.

P-6D

Army interest in the turbo-supercharger once again surfaced with the P-6D Hawk. The first P-6A (29-260) was re-engined at Wright Field with a Curtiss V-1570-23 engine fitted with a General Electric F-2E type turbo-supercharger under the designation XP-6D. The turbo-supercharger was mounted on the starboard side of the nose with large oil cooler air intake/exhaust ports on the port side. The exhaust system used a wrap-around exhaust manifold that channeled exhaust gasses from the port side of the engine to the supercharger on the starboard side. After a series of successful tests, the Army ordered nine P-6 aircraft and three P-6A aircraft converted to P-6D standards, with the actual conversion work being undertaken by U.S. Army Air Corps depots.

As with the earlier P-5, sea level performance fell off. The aircraft had a top speed of 172 mph and rate of climb of 1,720 ft/min. Once again the service ceiling was dramatically improved, topping out at 32,000 feet and the speed at 13,000 feet rose to 197 mph. With a maximum weight of some 3,483 pounds, the P-6D was the heaviest P-6 variant. The P-6D Hawk was also the first Hawk to be fitted with a Hamilton Standard three blade variable pitch propeller in place of the Curtiss two blade unit.

After service testing with the 37th Pursuit Squadron, 8th Pursuit Group at Langley Field, Virginia, a number of P-6Ds were used operationally with the 1st Pursuit Group at Selfridge Field as high altitude fighters. In any event, the turbo-supercharger proved to be unreliable and no further conversions or new production P-6Ds were produced. A number of P-6Ds remained in service until 1937, with the last aircraft being retired in April of 1938.

Fuselage Development

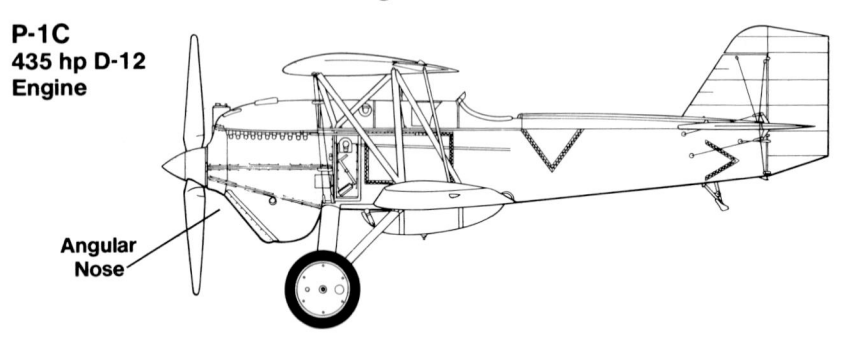

P-1C
435 hp D-12
Engine

Angular Nose

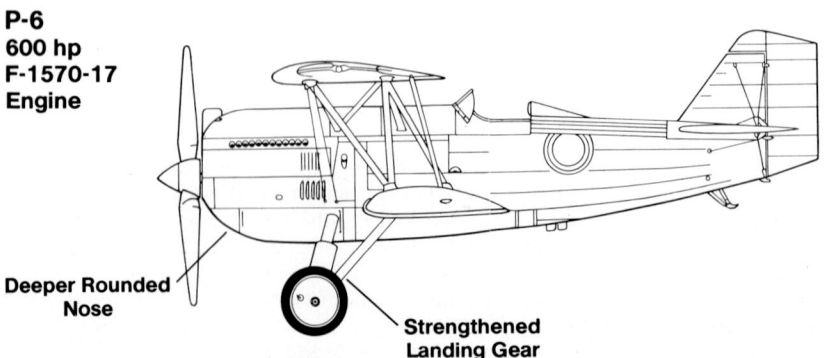

P-6
600 hp
F-1570-17
Engine

Deeper Rounded Nose

Strengthened Landing Gear

The pilot of this P-6 of the 36th Pursuit Squadron warms up the engine of his Hawk prior to starting another mission. The 36th PS was based at Selfridge Field during the Fall of 1930. The first nine P-6s used water cooled engines. (AFM)

The first P-6A Hawk (29-260) was assigned to the Fairfield Air Depot at Patterson Field near Dayton, Ohio. The P-6A used a standard P-6 airframe equipped with a smaller Prestone radiator. A total of nine P-6As were built. (Vincent Berinati)

The XP-6B was a purpose-built aircraft for use on long range record flights and had a 250 gallon fuselage fuel tank in place of the standard 50 gallon tank. CAPT Ross Hoyt flew the XP-6B from New York City to Nome, Alaska on 18/19 July 1929. (National Archives)

The heart of the XP-6D was the 600 hp Curtiss V-1570C Conqueror engine with its General Electric F-2E turbo-supercharger. The XP-6D differed from the P-6D in that it had a two blade propeller. (AFM)

Sixteen P-6 and P-6A aircraft were equipped with turbo-supercharged Curtiss V-1570-23 Conqueror engines driving three blade propellers under the designation P-6D. The supercharged engine brought the top speed at 15,000 feet to 197 mph and the service ceiling rose to over 31,000 feet. (Peter Bowers)

The P-6D shared the very rounded forward fuselage which was common to all early P-6 variants. This P-6D was assigned to the 37th PS at Langley Field, Virginia during 1934. The propeller spinner was Black and White, while the rest of the aircraft is the standard Army scheme. (Robert Cavanaugh)

A turbo-supercharged P-6D of the 37th Pursuit Squadron on the ramp at Langley Field, Virginia during 1934. The command stripe behind the cockpit, wheels and spinner were in Yellow, while the aircraft number was in White. (AFM)

Turbo Supercharger

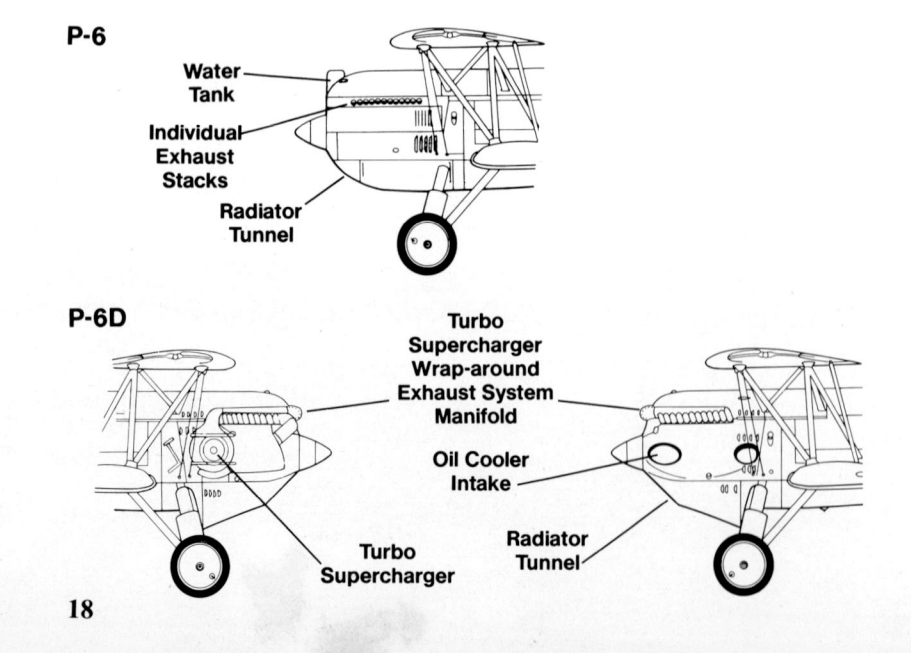

P-6
- Water Tank
- Individual Exhaust Stacks
- Radiator Tunnel

P-6D
- Turbo Supercharger Wrap-around Exhaust System Manifold
- Oil Cooler Intake
- Turbo Supercharger
- Radiator Tunnel

This Light Blue P-6D was assigned to the training detachment at Chanute Field, Illinois during the mid-1930s. Training units were often assigned older fighters as they were phased out of first line service. (Pete Bowers)

P-6E

The P-6E Hawk resulted from the mating of several features from different Curtiss experimental programs. It began in October of 1930 when one of the bulbous P-6 airframes was fitted with a 650 hp Wright R-1870-9 Cyclone radial engine in place of the standard Curtiss Conqueror under the designation YP-20. The aircraft featured standard P-6 oleo-pneumatic landing gear but the wheels were covered by aerodynamic wheel fairings known as "spats." The YP-20 also had an enlarged rudder, a spat-covered steerable tail wheel, improved cockpit layout and a hinged rear fuselage turtledeck to allow easy access to the baggage compartment and rear fuselage for maintenance. Overall performance with the Cyclone radial was slightly improved over the standard P-6, but not enough for the Army to justify putting the aircraft into production.

The second aircraft involved in development of the P-6E was the XP-22. The XP-22 used a P-6A airframe mated with a 600 hp V-1570-23 Conqueror engine housed in an entirely reshaped engine cowling. The much smaller Prestone radiator was moved to the rear and mounted in its own housing between a new style single-leg main landing gear. After a successful series of tests, the XP-22 was soon stripped of all its new features and put back into service as a production P-6A.

Taking the best features of the two experimental aircraft, Curtiss developed a new prototype under the designation XP-6E. The YP-20 fuselage was extensively modified with a 600 hp V-1570-23 Conqueror engine replacing the Wright Cyclone radial. The engine cowling was patterned after the one used with the XP-22, including the smaller Prestone radiator mounted between the single-leg landing gear with full wheel spats.

The aircraft carried the standard Army Air Corps fighter armament of two .30 caliber machine guns, but the guns were now mounted under the engine cylinder heads, firing through troughs in the engine cowling sides. Additionally, the XP-6E featured an unspatted steerable tail wheel, revised horizontal tail surfaces and three blade Hamilton Standard variable pitch propeller. Along with all these changes, the XP-6E also underwent a major weight reduction program and emerged some 412 pounds lighter than a standard P-6A (2,760 pounds).

The Army was very enthusiastic about the XP-6E and ordered forty-six production P-6Es in July of 1931. Performance tests with the P-6E revealed that the aircraft was superior in all ways to either the P-6A or the supercharged P-6D. The top speed was 198 mph at sea level, the rate of climb was 2,400 ft/min and the service ceiling was 24,700 feet. It compared favorably with its primary competition, the Boeing P-12. The P-6E was faster at all altitudes but slower in rate of climb and less maneuverable. The reason for this difference was the engine type. The Boeing P-12 was powered by an air-cooled Pratt & Whitney radial that was some 700 lbs lighter than the liquid cooled engine in the P-6E.

The first production P-6E was delivered to Wright Field for testing with the remainder of the production run being delivered to three squadrons: the 17th, 33rd and 95th Pursuit Squadrons. Of these only the 17th was solely equipped with P-6Es. These aircraft remained on active service through the mid-1930s, although accidents destroyed at least twenty-seven of the forty-six aircraft built. Most service aircraft had modified landing gear spats. The closed spats cause problems when operating off dirt and grass fields in wet weather or during the winter. Mud and or snow/ice would build up inside the spat interfering with the wheels. As a result, the outboard portion of the spat was removed, leaving the wheel uncovered. Finally, the last eighteen active P-6Es were retired from service between June and September of 1939. Most were donated to civil schools as non-flying training aids.

Three other P-6 variants were produced using P-6E airframes. The XP-6F was an XP-6E fitted with a turbo-supercharged V-1570F engine and a sliding cockpit canopy. Advances in supercharger technology allowed the XP-6F to attain a top speed of 225 mph at 15,000 feet. The aircraft remained active at Wright Field until October of 1937 when it was retired and donated to an aeronautical school.

The XP-6G was built from the 22nd production P-6E and served as an engine test bed for development work on the 700 hp V-1570F engine. Finally, the XP-6H was the first production P-6E with both the upper and lower wings modified to allow for the installation of a pair of .30 caliber guns in each wing, bringing the total armament to six .30 caliber guns. The aircraft was tested by the 8th Pursuit Group, 1st Pursuit Group and 17th Pursuit Group. With its increased armament, the XP-6H was the heaviest P-6 with a gross weight of 3,858 pounds. Its top speed was 190 mph and rate of climb was 2,000 ft/min. The air-

The Curtiss YP-20 was a P-11 (P-6 with Chieftain engine) re-engined with a 650 hp Wright R-1870-9 Cyclone air-cooled radial engine. The YP-20 airframe was later combined with certain features of the XP-22 to become the prototype XP-6E. (AFM)

The XP-22 was rebuilt from the third production P-6A. The modifications included single leg landing gear, a close-cowled engine and small Prestone radiator just in front of the main landing gear. (AFM)

craft was finally retired in July of 1939.

The final P-6E on the production contract became the XP-23 prototype and it bore little resemblance to any previous Hawk. The XP-23 had a completely new all-metal monocoque fuselage, which was very slender through the nose area, and a taller all-metal fin and rudder. The aircraft was powered by a 600-700 hp turbo-supercharged and geared G1V-1570-C engine. The wings and landing gear were all that remained from its P-6E heritage. Once again performance "at altitude" was outstanding, with a speed of 225 mph at 15,000 feet and a service ceiling of 33,000 feet. But rate of climb and sea level performance suffered. With the arrival of the monoplane fighter in the Boeing P-26, the Army Air Corps abandoned the XP-23 program. The XP-23 was finally disassembled and the wings and landing gear were used as the basis for the XF11C prototype for the Navy.

Fuselage Development

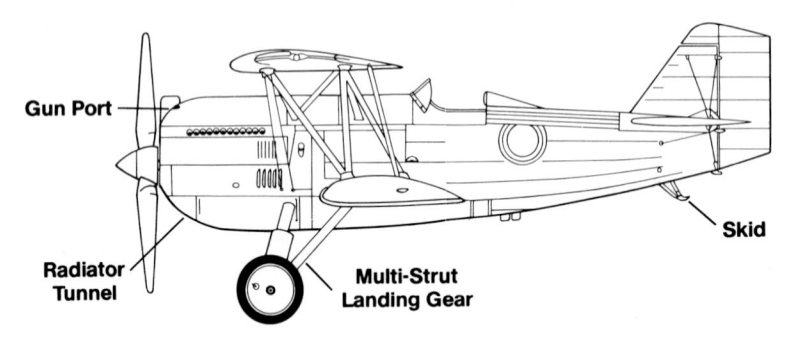

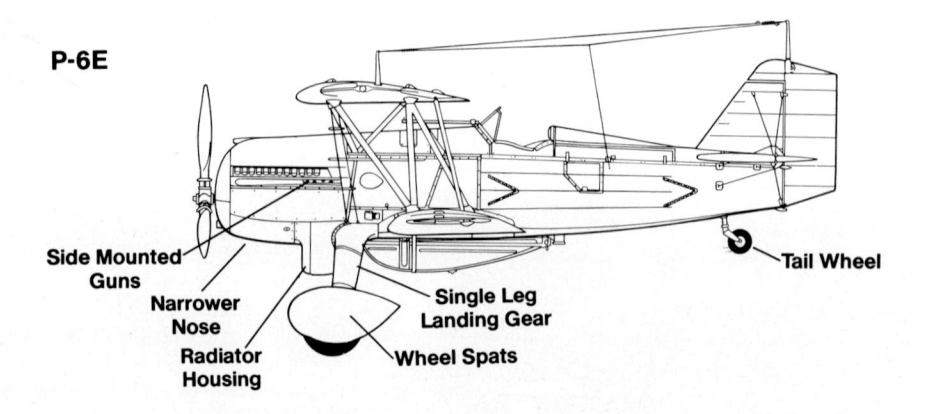

P-6

P-6E

The first production P-6E Hawk (32-233) sits on the ramp at Wright Field during late 1931. Although resembling the XP-22, the P-6E had a much slimmer nose section and had the guns relocated to the side of the nose. (Don Garrett, Jr.)

The cockpit of the P-6E was very simple. The large gauges are flight instruments, while the smaller gauges are engine monitors. The instrument panel had a Black leather finish. (AFM)

The two .30 caliber machine guns were mounted in the opening just below the engine exhaust stacks. The circular objects just behind the nose panel are the engine oil cooler assembly which fed cool air from the air scoop in the lower nose panel. (AFM)

The P-6E Hawk was powered by a 600 horse power Curtiss V-1570-23 Conqueror liquid cooled in-line engine. With a three blade Hamilton Standard variable pitch propeller, the P-6E Hawk had a top speed of 198 mph at sea level. (AFM)

This P-6E Hawk, of the 17th Pursuit Squadron at Selfridge Field, carries a underfuselage fifty-six gallon drop tank and has full wheel spats. This aircraft (32-264) was later sold to the Chicago Aero School during 1939 as a non-flying instructional airframe. (Fred Dickey)

A Curtiss P-6E Hawk of the 17th Pursuit Squadron, 1st Pursuit Group, based at Selfridge Field, Michigan during 1933. The Snow Owl markings were originally applied for the squadron's participation in the 1932 National Air Races. (AFM)

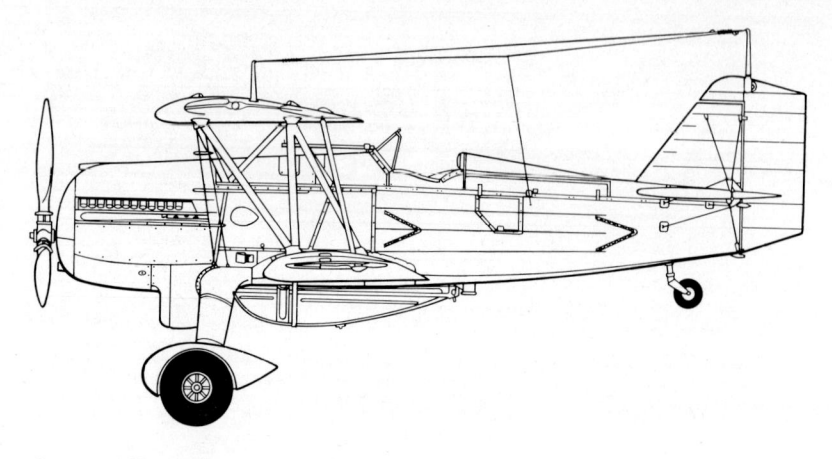

Specifications

Curtiss P-6E Hawk

Wingspan .... 31 feet 6 inches
Length .... 23 feet 2 inches
Height .... 8 feet 10 inches
Empty Weight .... 2,715 pounds
Maximum Weight .... 3,392 pounds
Powerplants .... One 600 hp Curtiss V-1570-23 Conqueror liquid cooled engine

Armament .... Two .30 caliber machine guns.

Performance
  Maximum Speed .... 198 mph
  Service ceiling .... 24,700 feet
  Range .... 244 miles
Crew .... One

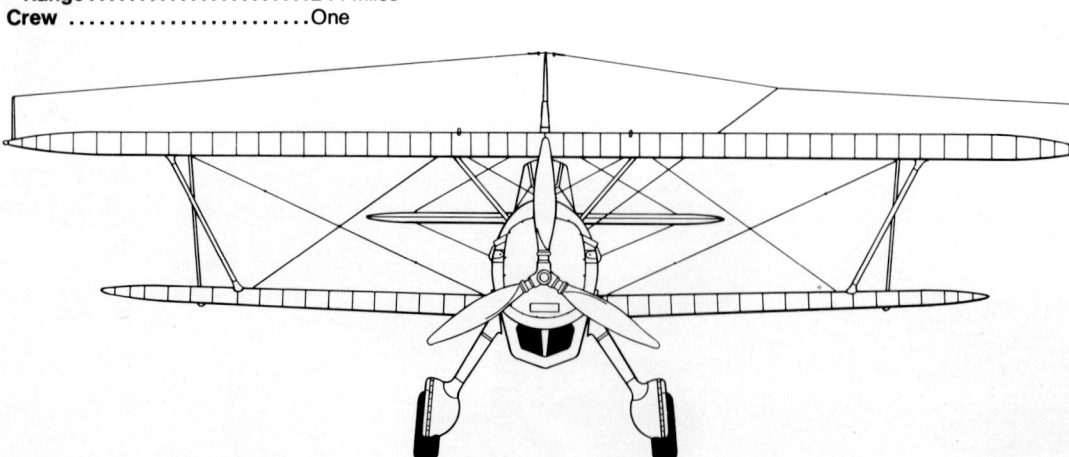

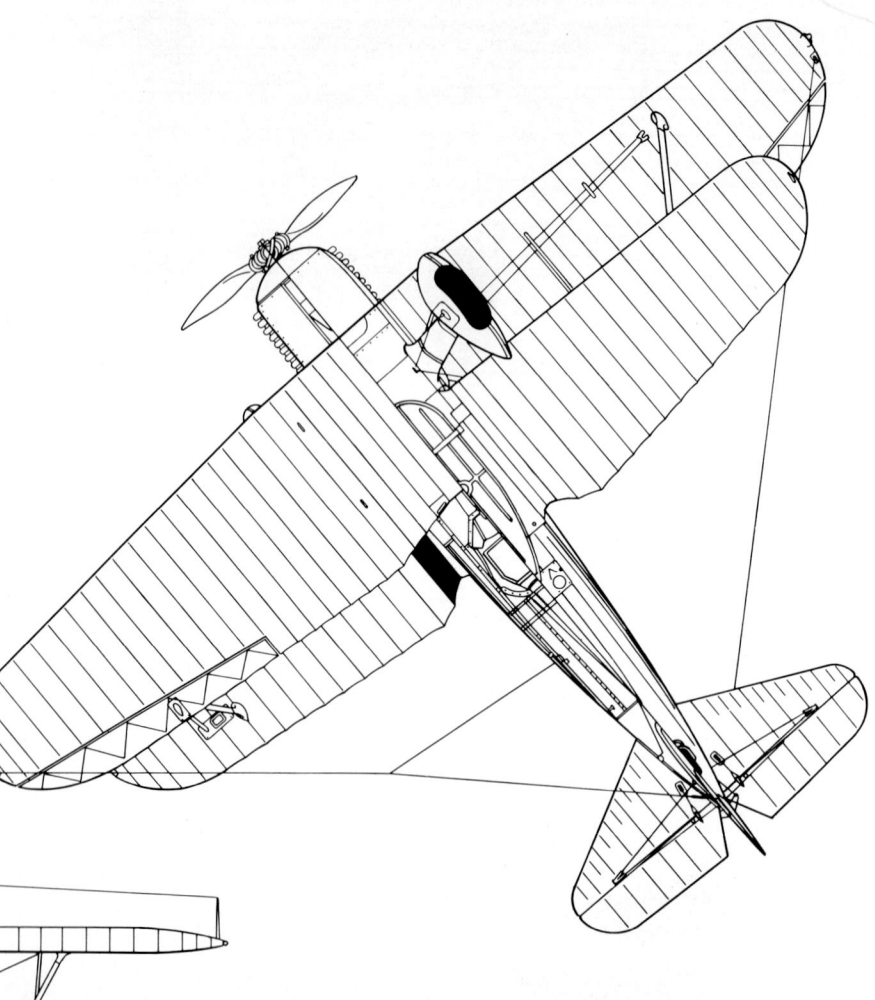

22

The P-6E had a larger fin and rudder than the earlier P-1 series and was equipped with a steerable tail wheel. The Snow Owl insignia of the 17th Pursuit Squadron differed from aircraft to aircraft within the unit. (AFM)

The P-6E carried some of the most colorful markings found on any pursuit aircraft. This P-6E (32-277) was assigned to the 94th PS at Selfridge Field during 1930. It had the standard Gloss Olive Drab fuselage with Yellow wings and tail surfaces, a Red/Yellow nose and the 94th unit insignia on both sides of the fuselage. (Fred Dickey)

In 1934 Army Air Corps standardized both Pursuit and Trainer aircraft color schemes. The new scheme had the fuselage painted in Light Blue #23 (FS 35109), with Yellow #4 wings and tail surfaces. This Wright Field P-6E carried the new scheme although the use of orthochromatic film made it appear to be overall Light Blue. (Jack Binder)

23

Two Army air groups operated the P-6E: the 1st Pursuit Group at Selfridge Field, Michigan and the 8th Pursuit Group at Langley Field, Virginia. This P-6E of the 8th PG has radio antennas mounted on the upper wing and vertical fin. (Fred Dickey)

P-6Es of the 17th Pursuit Squadron fly a line abreast formation during a squadron-strength, 2,840 mile distance record flight conducted during 1934. These aircraft carry the 17th's full Snow Owl markings which have become synonymous with the P-6E Hawk. (AFM)

A P-6E of the Bolling Field (Washington, D.C.) Detachment warms up its engine on the snow covered ramp at Floyd Bennett Field, New York during March of 1934. The fuselage number '1' indicates that the aircraft was assigned to the detachment commander. (Vincent Berinati)

U.S. ARMY

17th Pursuit Squadron

This P-1A Hawk (26-289), assigned to the 17th Pursuit Squadron at Selfridge Field during 1926, carried the early style USAAC rudder stripes.

P-1A Hawks of the 43rd School Squadron at Kelly Field, Texas during 1930 carried the revised rudder striping.

94th Pursuit Squadron

This Red nosed P-6E Hawk (32-241) flew with the 94th Pursuit Squadron at Selfridge Field, Michigan during 1934.

A Blue fuselage P-6E Hawk attached to Wright Field, Ohio as a station hack during 1937.

Wright Field

This 17th Pursuit Squadron P-6E Hawk (32-268) was outfitted with a ski landing gear during the Winter of 1933.

43rd School Squadron

Cuban Roundel

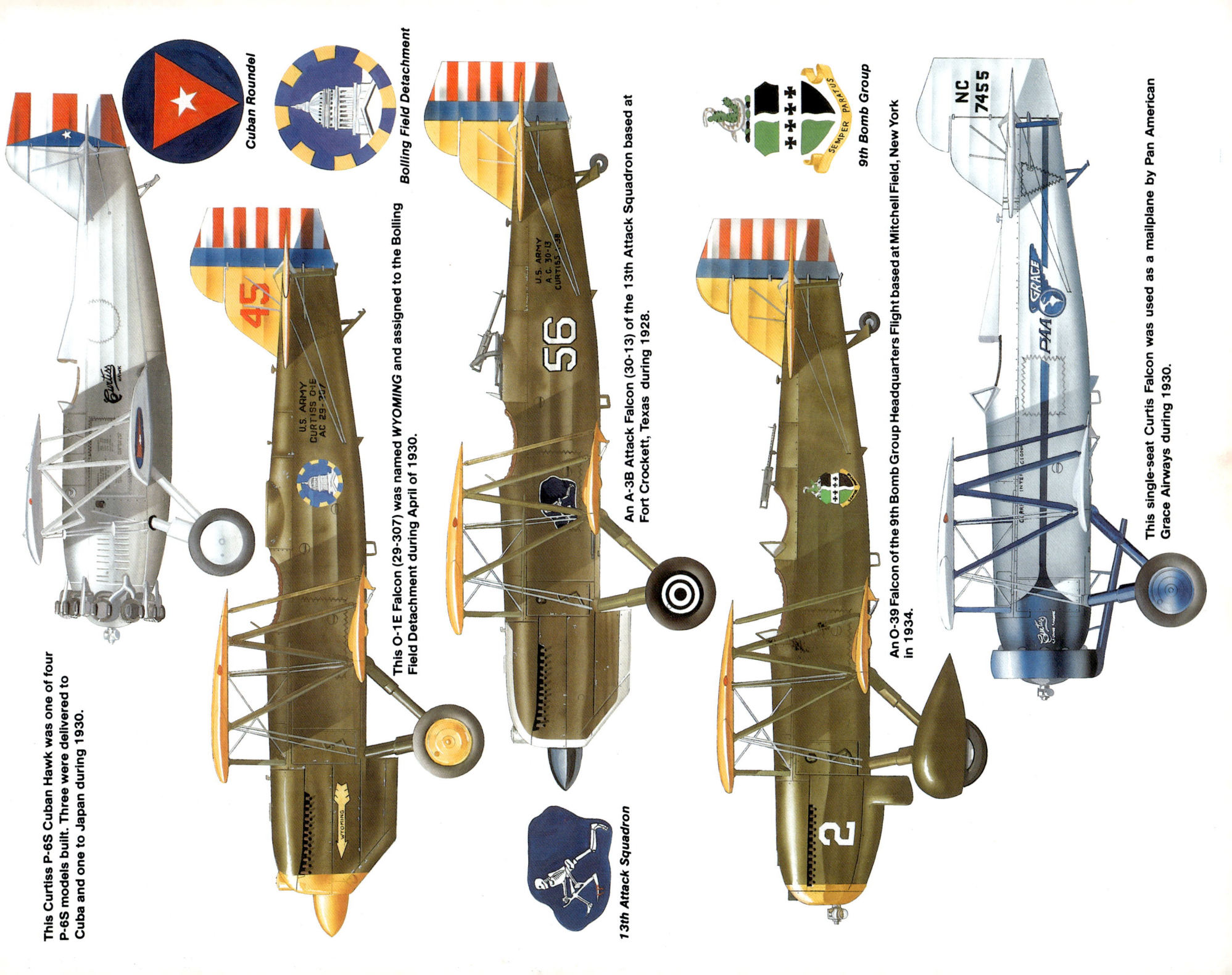

Bolling Field Detachment

9th Bomb Group

13th Attack Squadron

This Curtiss P-6S Cuban Hawk was one of four P-6S models built. Three were delivered to Cuba and one to Japan during 1930.

This O-1E Falcon (29-307) was named WYOMING and assigned to the Bolling Field Detachment during April of 1930.

An A-3B Attack Falcon (30-13) of the 13th Attack Squadron based at Fort Crockett, Texas during 1928.

An O-39 Falcon of the 9th Bomb Group Headquarters Flight based at Mitchell Field, New York in 1934.

This single-seat Curtis Falcon was used as a mailplane by Pan American Grace Airways during 1930.

This P-6E of the 1st PG carries a wargame camouflage scheme of Olive Green, Dark Green and Purple. This scheme was used during the 1933 Anti-Aircraft Exercises. Neutral Gray was substituted for Purple after it was found that the Purple paint actually attacked some fabric surface materials. (Peter Bowers)

A Yellow nosed P-6E Hawk of the 35th Pursuit Squadron P-6E at Langley Field, Virginia during May of 1937. The tail number on a P-6E usually, but not always, was the last two digits of the aircraft's serial number, in this case 32-241. (AFM)

This Light Blue P-6E from the 94th Pursuit Squadron carries the later designator code for the 1st PG "PA" and the individual aircraft number 98 (serial 32-298) on the fin, with the 94th Pursuit Squadron indian head insignia on the fuselage. Within a few months the 1st PG would transition to its first monoplane pursuit, the Seversky P-35. (Peter Bowers)

The XP-6F was a conversion of the original XP-6E (29-374) prototype re-engined with a 675 hp turbo-supercharged Curtiss V-1570-55 engine and a full cockpit canopy. The project did not progress beyond this single prototype. (AFM)

The XP-6H was a modification of the first production P-6E (32-233) with additional guns. A pair of .30 caliber machine guns were mounted in the upper wing and a second pair were mounted in the lower wing outside the propeller arc. The additional weight cut performance and the project was abandoned. (AFM)

The XP-6F was modified in June of 1934 with Type 8 ski landing gear which fitted over the standard wheeled landing gear and did not require removal of the wheels. The landing gear strut fairing and wheel spats were removed. (AFM)

The XP-23 was the ultimate biplane Hawk. It had an all-aluminum monocoque fuselage with taller tail surfaces. It was powered by a turbo-supercharged GIV-1570C engine, which gave it a top speed of 225 mph at 15,000 feet, making it the fastest of all the Hawks. (AFM)

Export Hawks

The large percentage of Curtiss export Hawks were based on Navy variants equipped with easier to maintain air cooled radial engines. Regardless, they were still simply known as Curtiss Hawks.

The first Hawk I export aircraft were standard P-1/P-1A/P-1B aircraft that had been approved by the Army for export sale. Seventeen P-1/Hawk Is were built, sixteen for the government of Chile and one for Japan. Additionally, eight standard P-6 aircraft were sold to the Dutch East Indies, with a further eight being built in Holland under license. These P-6s were also known as Hawk Is.

Several P-6 airframes were re-engined with 450 hp air cooled Pratt & Whitney Wasp radial engines under the designation P-6S. Other than the engine, they were identical to standard USAAC P-6s. Four aircraft were sold to Cuba during January of 1930 and one other was sold to Japan during the early 1930s.

Curtiss built one Hawk I for use as a company demonstrator aircraft which was often called the *Doolittle Hawk* thanks to the demonstrations put on in Europe and elsewhere by MAJ James Doolittle. Later the aircraft was sold to race pilot Jesse Bristow and registered as NX-9110. The aircraft was lost at sea in 1939 while enroute to Cuba.

The Hawk I-A was a purpose-built demonstrator aircraft flown by Al Williams and named *GULFHAWK*. The aircraft was a standard P-6 airframe with additional fuselage fuel tanks. Over its lifetime *GULFHAWK* used a variety of engines, including a standard V-1570 Conqueror and both Wright Cyclone and Bristol Jupiter air-cooled radial engines. The aircraft still exists and is on display at the Marine Corps Aviation Museum at Quantico, Virginia.

The export Hawk II series was based on the Navy F11C Goshawk design, which was basically a P-6E airframe re-engined with a 710 hp Wright R-1820F-3 radial engine and modified with naval equipment. Performance tests showed a top speed of 208 mph at 7,000 feet, with a service ceiling of 26,400 feet. Curtiss built some 127 Hawk II aircraft which were sold to Bolivia, Chile, China, Columbia (floatplanes), Cuba, Thailand, Turkey, Norway, and Nazi Germany. Several of the Hawk IIs saw combat in South American border conflicts and against the Japanese in the skies over China.

The Hawk III was the export version of the Navy BF2C-1 which had retractable landing gear, a semi-enclosed cockpit and wooden wings. The Hawk III was powered by a 750 hp Wright R-1820-53 Cyclone radial engine with a top speed of 240 mph and a service ceiling of 27,000 feet. Some 138 Hawk IIIs were produced and sold to Thailand, China, Turkey, and Argentina. Again, some saw combat especially over China against the invading Japanese. There was only one Hawk IV built, a demonstrator model with a fully enclosed cockpit, that was eventually sold to Argentina. Beyond the enclosed canopy, the Hawk IV was identical to the Hawk III and was the last of the biplane Hawk series. The day of the monoplane had arrived and with it an entirely new generation of Hawks.

MAJ Doolittle puts the the P-6 Hawk I demonstrator through its paces during the European tour. Because of his spirited demonstration flights the aircraft became known as the Doolittle Hawk. (AFM)

MAJ Jimmie Doolittle in the cockpit of the P-6 Hawk I demonstrator aircraft at Vienna, Austria during 1932. Sixteen Hawk I (P-6) aircraft were eventually sold to the Dutch East Indies and Royal Dutch Air Forces. (AFM)

The first export Hawks were P-1, P-1A, and P-1B aircraft cleared by the Army for export. The Curtiss Hawk I was based on the P-6 airframe with different power plants. This P-6 Hawk I was being shown to German officials during the 1932 Curtiss-Wright mission to Europe. (AFM)

Several Hawk Is were bought by civilians and used for various purposes. The Red, White, and Blue Esso Hawk I was flown by Jesse Bristow and used for flight demonstrations in the 1930s. The aircraft was lost at sea while enroute to Cuba for an air show. (Robert Cavanaugh)

The P-6S was a P-6 airframe fitted with a 450 hp Pratt & Whitney Wasp air cooled radial engine. A total of four P-6S aircraft were built, three being exported to Cuba as Cuban Hawks and the last to Japan.

Nose Development

P-6
600 hp Curtiss
V-1570A Engine

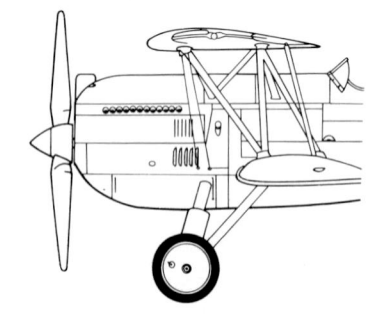

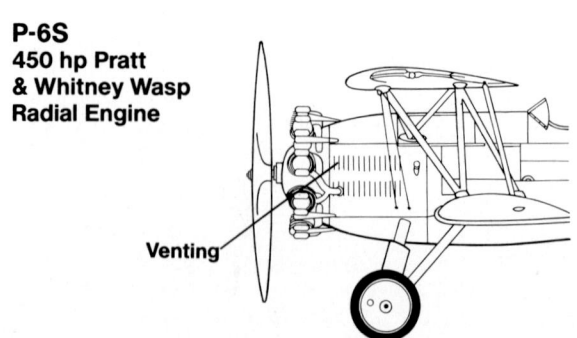

P-6S
450 hp Pratt
& Whitney Wasp
Radial Engine

Venting

Al Williams' first *GULFHAWK* was a specially built Hawk I-A powered by a 575 hp Wright Cyclone radial engine. The *GULFHAWK*, on the ramp at Roosevelt Field during 1931, was painted in traditional Gulf Oil Company colors of Orange and Dark Blue. (Vincent Berinati)

Al Williams modified the *GULFHAWK* a number of times. At Floyd Bennett Field during 1934, the *GULFHAWK* had an all-metal fuselage and was painted overall Gulf Oil Company Orange with Dark Blue trim. (Jack Binder)

The Navy and export variant of the P-6E Hawk was the XF11C-2 Goshawk. The aircraft was basically a P-6E powered by a 700 hp Wright R-1820 Cyclone radial engine and fitted with a arrester hook and relocated tail wheel. The top speed for the Goshawk/Hawk II was 202 mph. (Jeff Ethell)

The Curtiss Hawk III demonstrator (NR 14703) was based on the Navy BF2C-1 Goshawk which featured a retractable landing gear. The main difference between the Hawk III and Navy BF2C was that the Hawk III had wooden wings. (AFM)

Curtiss Falcon

The two seat Falcon series was a parallel development of the Hawk fighter series. Basically a larger Hawk, the Falcon shared the same bolted fuselage framework technology that was used with the Hawk. The Falcon used a PW-8/P-1 type fuselage that was extended 5.5 feet to accommodate a second cockpit for a rear gunner/observer. The wing planform differed from the Hawk in that the lower wing was not tapered and had a greater span than the lower wing on the P-1 Hawk. The upper wing was moved forward and had a straight center section with the outer wing panels swept back some nine degrees. The landing gear was the typical Curtiss Hawk style multi-strut main landing gear with a tail skid. Falcons were built as both observation and attack aircraft for the U.S. Army Air Corps.

O-1

The O-1 Falcon evolved from a 1924 Army competition for a new observation aircraft. Originally powered by the First World War Liberty engine, the XO-1 lost out badly to the Douglas XO-2. The Army, however, really did not want a Liberty engined aircraft so another competition was set for 1925 specifying the use of the new Packard 1A-1500 V-12 liquid cooled engine as the power plant. The Curtiss entry easily won and Curtiss was awarded a production contract for ten O-1 aircraft. Additionally, the specified power plant was changed from the Packard engine to the 435 hp Curtiss V-1150 (D-12). Production O-1s differed from the XO-1 in having an enlarged vertical fin and rudder. The armament consisted of a pair of flexible .30 caliber guns on a Scarf ring in the observer's cockpit and two forward firing .30 caliber guns mounted in the upper engine cowling.

The XO-1 Falcon observation aircraft rolled out in 1924 powered by a 510 hp Packard 1A-1500 liquid cooled engine. The O-1 Falcon was basically a larger Hawk, sharing the same fuselage design, landing gear and engine as the P-1 Hawk series. (AFM)

The O-1, known as the Falcon, had a gross weight of 4,400 pounds, almost 1,200 pounds more than a P-1 Hawk, although both used the same power plant. Performance tests showed a top speed of 136 mph, a rate of climb of 975 ft/min and a service ceiling of almost 16,000 feet. Deliveries of the ten production aircraft took place during 1925.

The O-1B was an O-1 airframe with the same improvements as the P-1C Hawk; larger steel wheels, mechanical brakes, a fuel jettison device for the main fuel tank and the ability to carry a fifty-six gallon under fuselage drop tank. The Army ordered forty-five O-1Bs, with deliveries starting in 1927. Four unarmed O-1Cs were built for specific use as VIP transport aircraft and were assigned to very high ranking officers in the Army, including the Secretary of the Army.

The O-1E differed from the O-1B/C in having the V-1150-5 engine, Frise ailerons, elevator balance horns, shock absorbers on the main landing gear and a refined cowl design. One O-1E was experimentally fitted with a cockpit canopy over the front cockpit. The Army ordered forty-one O-1Es during 1929, but several were modified into other prototype aircraft including the O-1F, XBT-4/XO-1G, YO-13C, and XO-26.

The O-1G had several notable changes over the O-1E. The gunner/observer cockpit was lowered into the fuselage and the sides of the fuselage cut out for easy entry. The scarf ring mount with its two .30 caliber machine guns was replaced by a vertical post mount carrying a single .30 caliber gun, the center of the upper wing was notched to improve the pilot's upward vision, the horizontal tail surfaces were increased in area and a steerable tail wheel replaced the earlier tail skid. The XO-1G prototype was built with full wheel spats, although this feature was deleted on production aircraft. The Army ordered a total of thirty production aircraft, which were delivered during 1931.

O-11

The next major variant was the O-11. The O-11 was built around an Army requirement for an observation aircraft for the National Guard. Production of the D-12 engine was barely enough to meet Army Air Corps pursuit aircraft needs so the Army decided to

The first production O-1 Falcon at the Curtiss plant in February 1926. The production O-1 differed from the prototype in that it was powered by a 600 hp Curtiss D-12 engine and had the height of the rudder increased. (AFM)

install 420 hp Liberty engines in O-1 airframes and issue these aircraft to National Guard units. The Liberty-engined O-1 prototype was designated the XO-11 and was followed by sixty-six production O-11s which used O-1B airframes powered by 435 hp Liberty 12A engines.

O-39

The O-39 was the final observation Falcon ordered into production. The O-39 used most of the same refinements found on the P-6E Hawk including the tightly-cowled V-1570-25 Conqueror engine and small Prestone radiator. The slimmer cowl lines and smaller radiator housing, plus the additional 165 horsepower, brought the top speed of the O-39 up to 173 mph. The rate of climb was also improved at 1,620 ft/min and the service ceiling rose to almost 23,000 feet. Production O-39s had P-6E style rudders, full wheel spats and multi-strut type landing gear. Armament was changed to a single .30 caliber machine gun in the cowling plus a flexible, post-mounted .30 caliber machine gun in the observer's cockpit. Although the O-39 was designed to use a full cockpit canopy and wheel spats, most aircraft in service had the canopy and spats deleted.

Fuselage Development

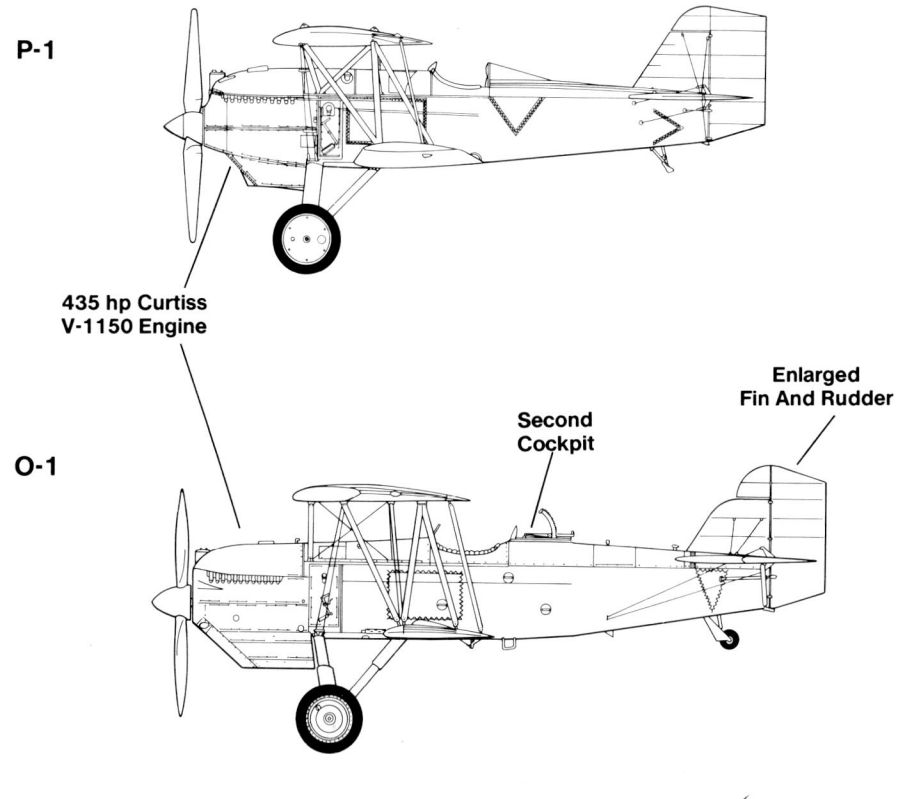

P-1

435 hp Curtiss V-1150 Engine

O-1

Second Cockpit

Enlarged Fin And Rudder

A-3 Attack Falcon

A natural development of the Falcon series was its modification for use as an attack aircraft. The first attack Falcon was an O-1B airframe modified with a pair of .30 caliber machine guns installed in the lower wings and underwing bomb racks. The aircraft was powered by the Curtiss V-1150-3 engine, giving the A-3 a top speed of 141 mph, a rate of climb of 1,046 ft/min and a service ceiling of almost 16,000 feet. Sixty-six A-3 Attack Falcons were ordered by the Army during 1927. Additionally, another six A-3s were built to fill the training role. These aircraft had the armament deleted and had dual flight controls. The Army ordered seventy-eight A-3B Attack Falcons during 1929. These aircraft had all the improvements found on the O-1E.

At least twenty Falcon airframes were built by Curtiss for civilian use as mailplanes and passenger aircraft. Most of these were powered by the 435 hp Liberty engine. National Air Transport bought fourteen single-seat mailplanes while Pan American Grace Airways had a single Falcon powered by a Wright Cyclone radial engine.

Export military sales amounted to a total of 135 aircraft. Fifteen O-1B aircraft were sold to Columbia and ten to Peru during 1928. These aircraft saw action in the war between the two countries that took place between 1932 and 1934. Ten O-1Es were built in Chile and some of these were transferred to Brazil where they saw combat during 1932 against rebel forces. At least 100 O-1G airframes powered by 712 hp Wright Cyclone engines were purchased by Columbia. These aircraft were used on both wheeled landing gear and floats.

A Curtiss O-1 parked on the grass at Wright Field. The radiator fairing has been removed showing the angle of the Heinrich radiator installation. The O-1 differed from the P-1 series in that the wing planform was changed with the upper wing, outer wing panels swept back from a straight center section. (AFM)

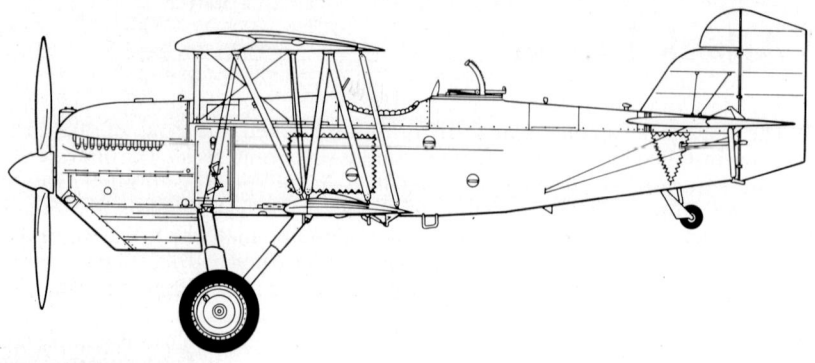

Specifications

Curtiss O-1B (A-3) Falcon

Wingspan . . . . 38 feet
Length . . . . 28 feet 4 inches
Height . . . . 10 feet 1 ½ inches
Empty Weight . . . . 2,227 (2,902) pounds
Maximum Weight . . . . 4,384 (4,458) pounds
Powerplant . . . . One 430 hp Curtiss V-1150 liquid cooled engine.

Armament . . . . Two fixed .30 caliber machine guns and two flexible .30 caliber guns in the observer's position (four fixed .30 caliber guns and 200 pounds of bombs for the A-3)

Performance
  Maximum Speed . . . . 135.5 mph (139 mph)
  Service ceiling . . . . 15,425 feet (14,100 feet)
  Range . . . . 595 miles (647 miles)
Crew . . . . Two

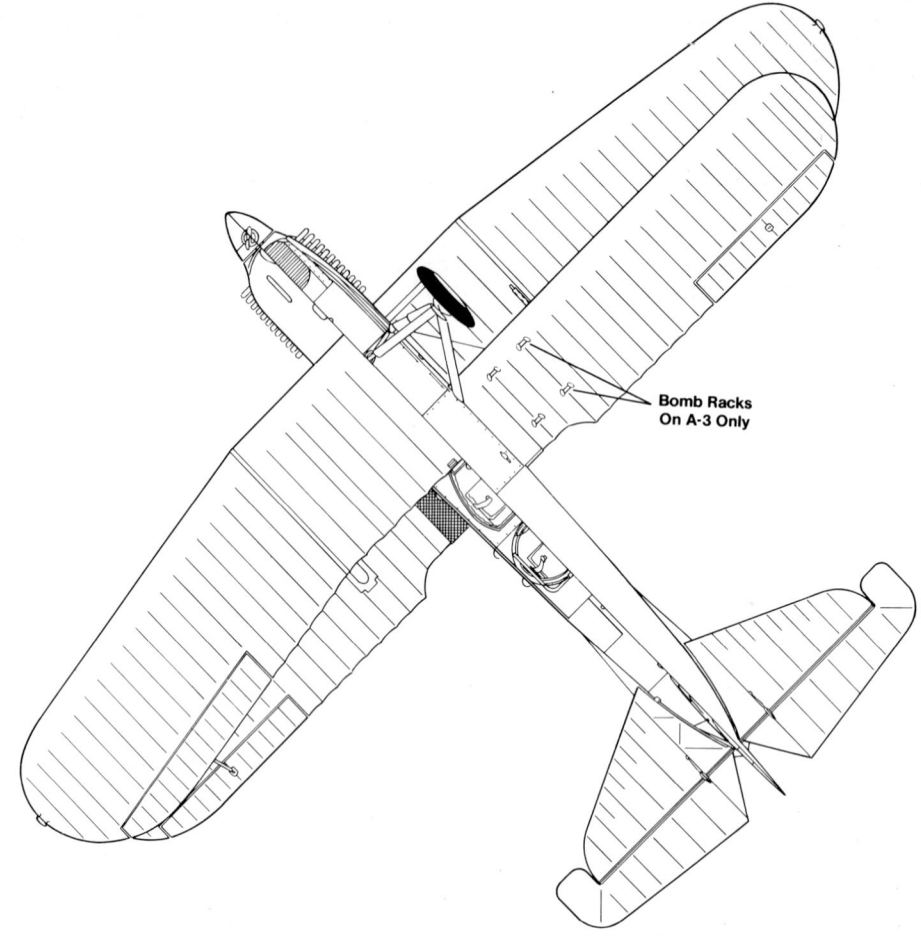

Bomb Racks On A-3 Only

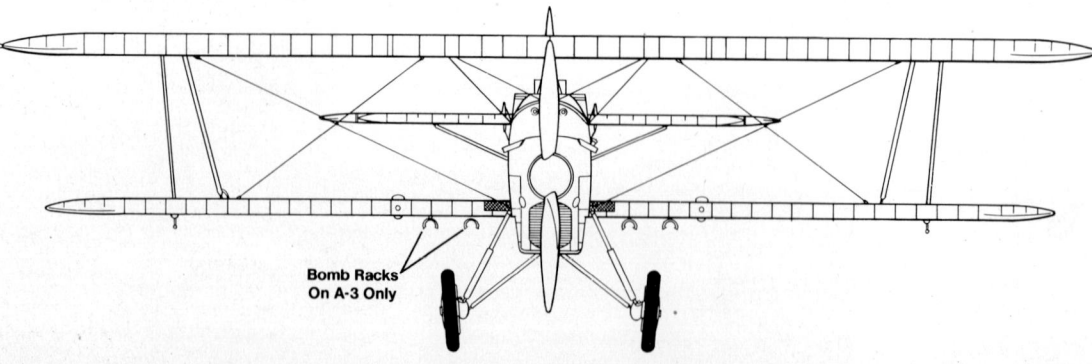

Bomb Racks On A-3 Only

36

The pilot's cockpit of the O-1 Falcon was Natural Metal while the instrument panel was Black Leather. The control quadrant in the center had labeled control levers for the throttle (T), propeller pitch (P) and mixture (M). (AFM)

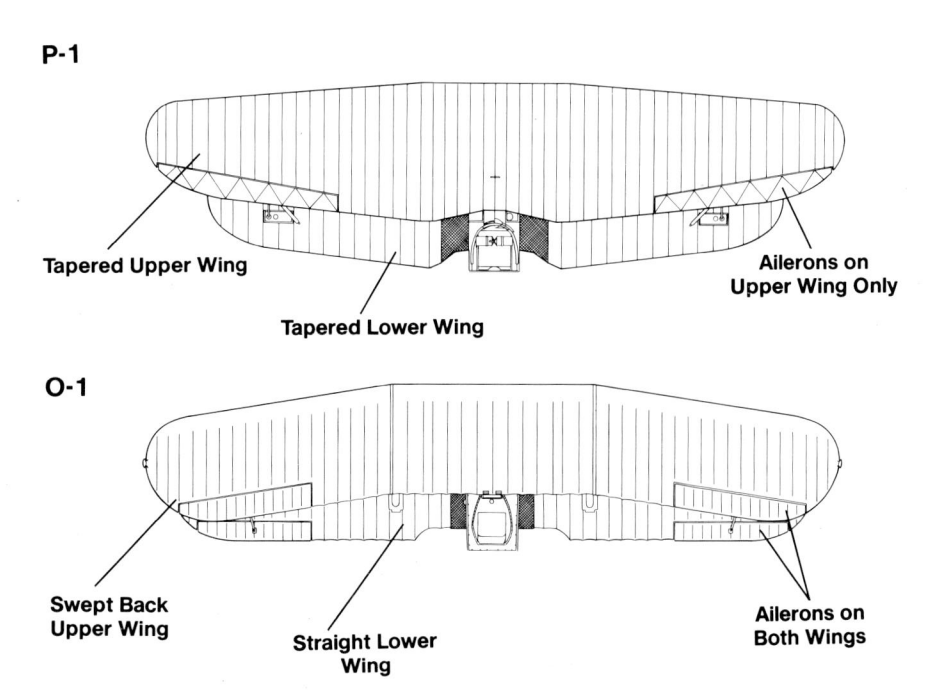

This O-1 was assigned to the Air Corps Technical School at Kelly Field, Texas. The O-1 had a single fixed .30 caliber gun in the engine cowling, plus a flexible .30 caliber gun mounted on the Scarf ring for the observer. This O-1 has been retrofitted with a tail wheel in place of the tail skid. (AFM)

Wing Development

P-1

- Tapered Upper Wing
- Tapered Lower Wing
- Ailerons on Upper Wing Only

O-1

- Swept Back Upper Wing
- Straight Lower Wing
- Ailerons on Both Wings

One O-1 was re-engined with a Liberty engine under the designation O-1A. The aircraft was taken into service with the serial 25-333 and assigned to a USAAC observation squadron where it received the name VULCAN. (Warren Bodie)

The O-1B Falcon featured the same improvements as the P-1C Hawk, including the installation of mechanical brakes and the fifty-six gallon under fuselage fuel drop tank. Curtiss built a total of forty-five O-1Bs during 1927/28. (AFM)

Five O-1Bs from the Air Corps Bolling Field Detachment share the Bolling Field ramp with MGEN Benjamin Foulois' Douglas O-2D. MGEN Foulois was the Chief of the Army Air Corps during 1932. (AFM)

This O-1C was assigned to the Secretary of the Army, Mr. Davison (in rear cockpit), during 1927. The aircraft is Gloss Olive Drab with Yellow wings and tail surfaces. The OD painted fabric areas of the fuselage appear lighter in color than the metal areas, which are Gloss Black. (AFM)

This O-1C was assigned to MGEN Fechet, the outgoing Chief of the Army Air Corps. The two stars on the fin identified the aircraft as being assigned to a General Officer. The aircraft was overall Gloss Olive Drab with a Gloss Black cowling, upper fuselage and fin band. (AFM)

One O-1E was fitted with a fully enclosed forward cockpit canopy and assigned to the Middletown Air Depot at Wright Field for testing. The depot marking M.A.D. was painted on the rudder in Black. (AFM)

This O-1E of the Air Corps Bolling Field Detachment was named *WYOMING*. The nose and arrow were in Yellow and the aircraft number in the tail appears to be in Red. The O-1E introduced Frise ailerons and shock absorbers on the main landing gear. (AFM)

39

An O-1G of the 9th Bomb Group at the National Air Races during 1929. Production O-1Gs had the wheel spats deleted. The number 2 on the fuselage is in White while the number under the wing is in Gloss Black, not White (reflected light off the ground makes it appear lighter). (Jack Binder)

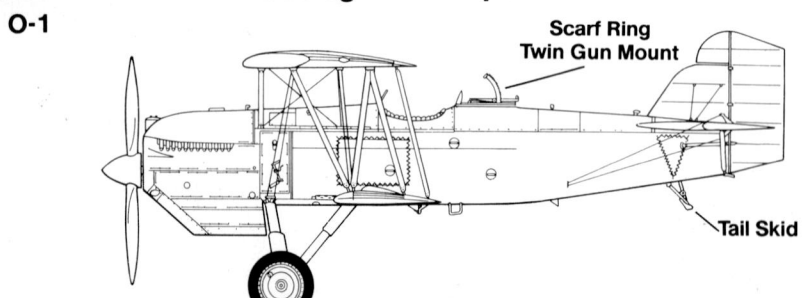

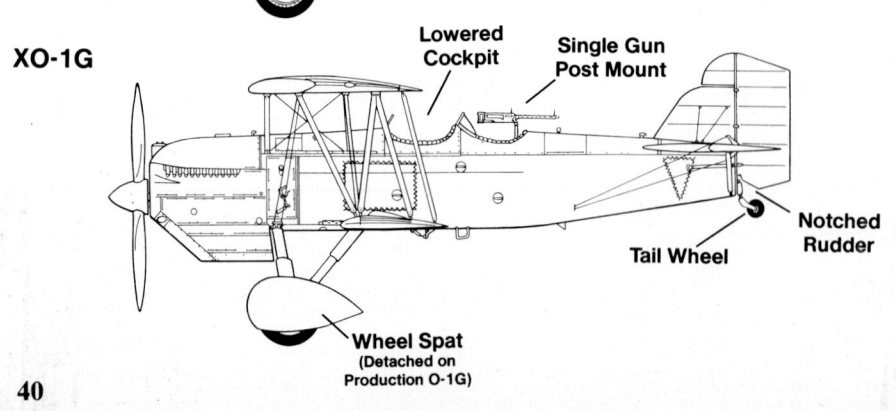

The XO-1G prototype on the ramp at Wright Field. The XO-1G had the gunner/observer cockpit lowered into the fuselage and the gun mount was changed from the Scarf ring twin gun mount to a single gun post mount. Additionally, the aircraft was fitted with a steerable tail wheel. (Robert L Cavanaugh)

Fuselage Development

O-1
- Scarf Ring Twin Gun Mount
- Tail Skid

XO-1G
- Lowered Cockpit
- Single Gun Post Mount
- Tail Wheel
- Notched Rudder
- Wheel Spat (Detached on Production O-1G)

40

Several O-1Gs and a DH-4B on the grass at Fort Ethan Allen, Vermont. The aircraft were assigned to the Photo Section, 18th Brigade during September of 1928. Curtiss built a total of thirty O-1Gs. The aircraft in the foreground with the star markings belonged to the Commander of the First Army. (AFM)

A line-up of O-11 Falcons at Kelly Field, Texas in December of 1931. The Liberty-engined O-11 was used as a transition trainer aircraft during the early 1930s, as well as being the primary observation aircraft of the National Guard. The aircraft are equipped with wire and mesh stone guards over the wheels. (AFM)

This O-1G is painted with the Light Blue #23 fuselage with Yellow wings that was standardized by the Army during 1934. By this date most of the O-1s remaining in service had been relegated to the trainer role or were used as unit hacks. (R. Volker)

An O-1G Falcon of the 97th Observation Squadron at Mitchell Field, New York during 1930. The aircraft has a landing light installed on the leading edges of both lower wings. (Peter Bowers via Steve Hudek)

Pilots from the Photo Section of the 106th Observation Squadron, Alabama National Guard, pose in front of an O-11 at Roberts Field, Birmingham, Alabama. Curtiss built a total of sixty-six O-11s for issue to National Guard units. (AFM)

The paint schemes for National Guard aircraft were also changed by the 1934 directive that changed pursuit and trainer aircraft fuselages from Olive Drab to Light Blue #23. This O-11 also has been retrofitted with a steerable tail wheel. (Fred Dickey)

LT. Steensen flew this O-11 Falcon during the late 1920s. The O-11 used an O-1B airframe powered by a 420 hp Liberty engine. O-11s were built solely for use by National Guard forces. (Steenson Collection)

The XO-13 was painted overall Olive Drab including the wing and tail surfaces and was powered by a Curtiss V-1570 Conqueror engine. The aircraft placed Second in the two-place category in the 1927 National Air Race. (AFM)

The XO-13A used a Conqueror engine equipped with PW-8 type, flush-mount radiators in the upper and lower wings. The XO-13A won the two-seat class at the 1927 National Air Race, at an average speed of over 170 mph. (AFM)

The A-3 was an attack version of the O-1 Falcon observation aircraft. It was equipped with underwing bomb racks and five .30 caliber guns: two in the lower wings, two in the engine cowling and one in the observer's cockpit. (AFM)

A pair of O-13Cs from the 5th Observation Squadron, 9th BG at Mitchell Field. Three O-13Cs were built as service test aircraft. The O-13 was basically an O-1E airframe mated with a 600 hp Conqueror engine. (Robert L. Cavanaugh)

An A-3A of the Air Corps Technical School at Kelly Field, Texas during 1928. The A-3A could be used for both the attack and observation roles since it could be fitted with a camera behind the gunner's cockpit. (AFM)

An A-3B from the 13th Attack Squadron at Fort Crockett, Texas during May of 1933. The A-3 had bomb racks under the wings and was capable of carrying up to 200 pound of bombs. (AFM)

In addition to its fixed forward firing armament, the A-3 was armed with a single flexible .30 caliber machine gun on a Scarf ring mount for the observer. (George Fisher)

An A-3A from the 41st Reconnaissance Squadron during June of 1930. By this date most A-3s had been stripped of their forward firing armament and bomb racks and were assigned to the observation squadrons, serving along the O-1s remaining in service. (AFM)

This A-3B carried the markings of the 90th Attack Squadron. The rear gun has been deleted and a tall radio antenna mast has been added to the rear fuselage indicating that this A-3 is serving as a command aircraft. (AFM)

This A-3B was assigned to the Training School at Chanute Field, Illinois. The winter weather conditions at northern fields weathered the paint very rapidly giving this aircraft a patchy appearance. This A-3B has been stripped of armament and was being used as a trainer. (Robert Esposito)

An A-3B from the 13th Attack Squadron at Fort Crockett, Texas. This aircraft has been modified with underwing beacon lamps for night flying. The aircraft was Olive Drab with Yellow wings and tail surfaces, and has a White nose, Black spinner and Black and White wheels. (Warren Bodie)

The 3rd Attack Group lined up on the grass at Fort Crockett, Texas during the late 1920s. The 3rd Attack Group had four squadrons: the 8th, 13th, 26th and 90th Attack Squadrons, all equipped with A-3B aircraft, the only Air Corps group that was solely equipped with the A-3. (AFM)

One O-1B was modified to use a 440 hp Pratt & Whitney Wasp air cooled radial engine under the designation XA-4. The US Navy purchased twenty-five aircraft based on the XA-4 proposal for the Marine Corps, designating them F8C-1/-3. (AFM)

The YO-1G began life as an O-1E that was re-equipped with a post gun mount in the observer's cockpit and a steerable tail wheel. Thirty production O-1Gs were based on the Y1O-1G. Later the YO-1G was re-engined with a V-1570 engine and small radiator. (Warren Bodie)

This O-1G was assigned to Patterson Field in September 1937. It has been modified with a full canopy fitted over both cockpits and has been re-engined with a V-1570 Conqueror engine and small Prestone radiator. (Steven Hudek)

The XO-39 Falcon observation aircraft was the final development in the Army Hawk/Falcon line. The XO-39 was powered by the Curtiss V-1570 Conqueror engine with the smaller Prestone radiator as used on the P-6E Hawk. (AFM)

A Wright Field O-39 parked on the grass at the Middletown Air Depot. Production O-39s had all the improvements found on the O-1G Falcon, but were powered by the V-1570 Conqueror engine. (AFM)

An O-39 of the 99th Observation Squadron at Mitchell Field. The O-39s were delivered with full wheel spats, which were usually removed due to landing ground conditions. The deeply cut back fuselage sides in the observer's cockpit area was common to the O-1G and the O-39. (Warren Bodie)

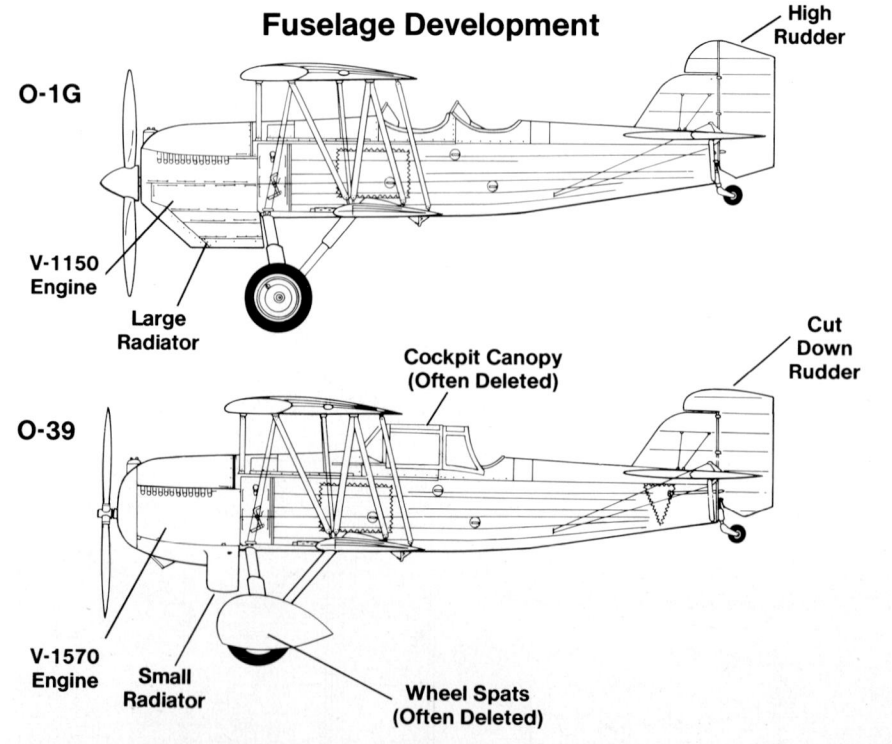

An O-39 from the Headquarters Detachment, 9th BG at Mitchell Field, shares the ramp with a 9th BG Douglas Y1B-7. The O-39 has an Olive Drab fuselage with Yellow wings, nose and tail. The aircraft is equipped with landing lights on the wing leading edge. (AFM)

Fuselage Development

O-1G
- V-1150 Engine
- Large Radiator
- High Rudder

O-39
- V-1570 Engine
- Small Radiator
- Cockpit Canopy (Often Deleted)
- Wheel Spats (Often Deleted)
- Cut Down Rudder

This Curtiss Falcon mail plane was powered by a 745 hp Wright Cyclone air-cooled radial engine. The aircraft was used by Pan American Airways, Grace Division in Latin America. (AFM)

This Falcon was used by the Wright Aeronautical Corporation as a test bed for geared and turbo-supercharged Conqueror engine experimental work. The aircraft was originally built as a Liberty powered aircraft. (Vincent Berinati)

This O-39 of the 97th Observation Squadron was repainted with a Light Blue #23 fuselage in accordance with the 1934 directive. The aircraft is equipped with a radio antenna array on the upper wing and fuselage and had navigational lights added to the wing and tail tips. (Warren Bodie)

Most export versions of the Falcon were powered by Wright Cyclone radial engines, including those sold to Columbia in 1932. The aircraft were equipped with floats and saw combat during the 1932-34 war between Peru and Columbia. (Peter Bowers via Steven Hudek)

Army Air Force in Action

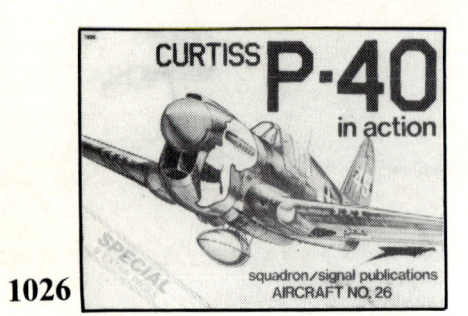

1026 — Curtiss P-40 in action
1034 — B-25 Mitchell in action
1045 — P-51 Mustang in action
1063 — B-17 in action
1067 — P-47 Thunderbolt in action
1080 — B-24 Liberator in action
1106 — P-61 Black Widow in action
1109 — P-38 Lightning in action

squadron/signal publications